GUN CONTROL

ISSN 1534-1909

GUN CONTROL

Mark Lane

INFORMATION PLUS® REFERENCE SERIES
Formerly Published by Information Plus, Wylie, Texas

GALE
CENGAGE Learning·

Farmington Hills, Mich • San Francisco • New York • Waterville, Maine
Meriden, Conn • Mason, Ohio • Chicago

GALE
CENGAGE Learning·

Gun Control

Mark Lane
Kepos Media, Inc.: Steven Long and
Janice Jorgensen, Series Editors

Project Editors: Laura Avery, Tracie Moy

Rights Acquisition and Management: Ashley
  Maynard

Composition: Evi Abou-El-Seoud, Mary Beth
  Trimper

Manufacturing: Rita Wimberley

© 2015 Gale, Cengage Learning

For product information and technology assistance, contact us at
Gale Customer Support, 1-800-877-4253.
For permission to use material from this text or product,
submit all requests online at **www.cengage.com/permissions.**
Further permissions questions can be e-mailed to
permissionrequest@cengage.com

Cover photograph: © Stephanie Frey/Shutterstock.com.

Gale
27500 Drake Rd.
Farmington Hills, MI 48331-3535

ISBN-13: 978-0-7876-5103-9 (set)
ISBN-13: 978-1-57302-642-0

ISSN 1534-1909

This title is also available as an e-book.
ISBN-13: 978-1-57302-676-5 (set)
Contact your Gale sales representative for ordering information.

Printed in the United States of America
1 2 3 4 5   19 18 17 16 15

TABLE OF CONTENTS

PREFACE

Gun Control is part of the *Information Plus Reference Series*. The purpose of each volume of the series is to present the latest facts on a topic of pressing concern in modern American life. These topics include the most controversial and studied social issues of the 21st century: abortion, capital punishment, child abuse, crime, the economy, energy, health care, immigration, national security, social welfare, women, youth, and many more. Although this series is written especially for high school and undergraduate students, it is an excellent resource for anyone in need of factual information on current affairs.

By presenting the facts, it is the intention of Gale, Cengage Learning, to provide its readers with everything they need to reach an informed opinion on current issues. To that end, there is a particular emphasis in this series on the presentation of scientific studies, surveys, and statistics. These data are generally presented in the form of tables, charts, and other graphics placed within the text of each book. Every graphic is directly referred to and carefully explained in the text. The source of each graphic is presented within the graphic itself. The data used in these graphics are drawn from the most reputable and reliable sources, such as from the various branches of the U.S. government and from private organizations and associations. Every effort has been made to secure the most recent information available. Readers should bear in mind that many major studies take years to conduct and that additional years often pass before the data from these studies are made available to the public. Therefore, in many cases the most recent information available in 2015 is dated from 2012 or 2013. Older statistics are sometimes presented as well, if they are landmark studies or of particular interest and no more-recent information exists.

Although statistics are a major focus of the *Information Plus Reference Series*, they are by no means its only content. Each book also presents the widely held positions and important ideas that shape how the book's subject is discussed in the United States. These positions are explained in detail and, where possible, in the words of their proponents. Some of the other material to be found in these books includes historical background, descriptions of major events related to the subject, relevant laws and court cases, and examples of how these issues play out in American life. Some books also feature primary documents or have pro and con debate sections that provide the words and opinions of prominent Americans on both sides of a controversial topic. All material is presented in an evenhanded and unbiased manner; readers will never be encouraged to accept one view of an issue over another.

HOW TO USE THIS BOOK

Gun control is a topic that elicits passionate emotions. On one side are those who believe the right to own guns should be subject to more stringent restrictions and regulations, given their propensity to be used to commit crimes and suicide. This side tends to see firearm violence as a public health issue and to advocate the kind of robust government response that an epidemic of a disease might elicit. On the other side are those who consider the Second Amendment to be one of the quintessential American freedoms. This side tends to see restrictions on the right to own guns as an infringement of basic liberty and as progress down a slippery slope that will end in the banning of all firearms.

Gun Control consists of 10 chapters and four appendixes. Each chapter is devoted to a particular aspect of gun control in the United States. For a summary of the information that is covered in each chapter, please see the synopses that are provided in the Table of Contents. Chapters generally begin with an overview of the basic facts and background information on the chapter's topic, then proceed to examine subtopics of particular interest. For example, Chapter 6: Gun-Related Injuries and Fatalities begins by explaining the current sources of the best government data on firearm injuries and fatalities and noting legal restrictions on the collection and dissemination of such data. The chapter then proceeds to present statistics relating to nonfatal firearm injuries, firearm fatalities, and the costs to society of firearm injuries. Other topics covered include a presentation of dueling estimates of the number of firearm injuries that occur as a result of lawful self-defense, and an overview of prominent gun safety laws and initiatives. Readers can find their way through a chapter by looking for the section and subsection headings, which are clearly set off from the text. They can also refer to the book's extensive Index if they already know what they are looking for.

Statistical Information

The tables and figures featured throughout *Gun Control* will be of particular use to readers in learning about this issue. These tables and figures represent an extensive collection of the most recent and important statistics on gun ownership, gun violence, and gun control legislation, as well as the trends—for example, graphics outline the percentage of Americans who own various types of firearms, the number and result of criminal background checks under the Brady gun control laws, and the decline in the firearm violence rate that has occurred since the mid-1990s. Gale, Cengage Learning, believes that making this information available to readers is the most important way to fulfill the goal of this book: to help readers understand the issues and controversies surrounding gun control in the United States and reach their own conclusions.

Each table or figure has a unique identifier appearing above it, for ease of identification and reference. Titles for the tables and figures explain their purpose. At the end of each table or figure, the original source of the data is provided.

To help readers understand these often complicated statistics, all tables and figures are explained in the text. References in the text direct readers to the relevant statistics. Furthermore, the contents of all tables and figures are fully indexed. Please see the opening section of the Index at the back of this volume for a description of how to find tables and figures within it.

Appendixes

Besides the main body text and images, *Gun Control* has four appendixes. The first appendix lists the 50 states and cites the specific articles of the 44 states that have constitutional provisions regarding weapons. The second appendix is the Important Names and Addresses directory. Here, readers will find contact information for a number of government and private organizations that can provide further information on aspects of gun control. The third appendix is the Resources section, which can also assist readers in conducting their own research. In this section, the author and editors of *Gun Control* describe some of the sources that were most useful during the compilation of this book. The final appendix is the Index. It has been greatly expanded from previous editions and should make it even easier to find specific topics in this book.

COMMENTS AND SUGGESTIONS

The editors of the *Information Plus Reference Series* welcome your feedback on *Gun Control* Please direct all correspondence to:

Editors
Information Plus Reference Series
27500 Drake Rd.
Farmington Hills, MI 48331-3535

CHAPTER 1
THE HISTORY OF THE RIGHT TO BEAR ARMS

The right to bear arms has deep roots in American legal and cultural traditions. From the time colonists settled on North American soil, Americans have held weapons to protect themselves. Armed citizen-soldiers won America's freedom from English rule more than two centuries ago. Partly because of this connection to the country's history, attempts to restrict a citizen's right to own a gun evoke strong emotions.

The modern debate over gun control erupted after a series of high-profile assassinations during the 1960s, including the assassinations of President John F. Kennedy (1917–1963), the civil rights leader Martin Luther King Jr. (1929–1968), and Senator Robert F. Kennedy (1925–1968; D-NY). In the decades that followed, the debate gained new urgency as gun-related violence among ordinary Americans increased. In the regularly updated series *Homicide Trends in the United States* (2014, http://www.bjs.gov/index.cfm?ty=pbse&sid=31), the Bureau of Justice Statistics indicates that gun-related homicides rose sharply in the late 1970s and peaked in 1980 and that the number of homicides rose again beginning in the late 1980s and peaked in the mid-1990s. Although the annual numbers of gun-related homicides began dropping nationally after the mid-1990s' peak, mass shootings such as the 2007 gun rampage at Virginia Polytechnic Institute and State University and the 2012 killings of 20 first graders and six staff members at Sandy Hook Elementary School in Connecticut have kept the issue of gun control in the forefront of the collective American consciousness.

At the heart of the gun control debate is the interpretation of the Second Amendment to the U.S. Constitution. One side claims that gun ownership is an individual right guaranteed by the Second Amendment and that guns are vital for self-protection. The other side believes guns should be banned or restricted because many innocent lives are lost due to their misuse. Gun control advocates say there is no longer a compelling need for people

to "keep and bear arms" as there was when the Constitution was ratified in 1788. They sometimes add the argument that the constitutionally guaranteed right was never meant to apply to individuals. This argument was supported by a consensus of legal experts and the U.S. Supreme Court until the 21st century, when an interpretation focusing on the individual right to bear arms won a majority of adherents among Supreme Court justices. Regardless of the merits of either side's argument or of the interpretations of legal experts, polls show that a majority (68%) of Americans understand that the Second Amendment authorizes an individual's right to own firearms. (See Table 1.1.)

The right of the individual to keep and use weapons has a long tradition in Western civilization. The Greek philosopher Aristotle (384–322 BC) wrote in *Politics* that ownership of weapons was necessary for true citizenship and participation in the political system. By contrast, another Greek philosopher, Plato (428–348 BC), wrote in the *Republic* that he believed in a monarchy with few liberties and saw the disarming of the populace as essential to the maintenance of an orderly and autocratic system. In *De Officiis*, the Roman politician Marcus Tullius Cicero (106–43 BC) expressed support for the individual's right to own weapons for self-defense and for public defense against tyranny. Similarly, in *Discourse*, the Italian political philosopher Niccolò Machiavelli (1469–1527) advocated an armed populace of citizen-soldiers to keep headstrong rulers in line.

AN EARLY PRECEDENT: MILITIAS AND THE OWNERSHIP OF WEAPONS

One of the first documents to link the bearing of arms with a militia (an army composed of citizens called to action in time of emergency) was the English *Assize of Arms* of 1181, which directed every free man to have access to weaponry. Henry II of England (1133–1189)

TABLE 1.1

Poll respondents' views on the meaning of the Second Amendment, 2014

"Please read the following passage: 'A well regulated militia, being necessary to the security of a free state, the right of the people to keep and bear arms, shall not be infringed.' Now, for each of the following, please indicate whether you feel the passage does or does not support it."

Base: U.S. adults		Supports this	Does not support this
A state militia's right to own firearms	%	74	26
Any citizen's right to own firearms	%	68	32
A state's right to form a militia	%	65	35
A state's right to regulate the ownership of firearms among its militia	%	55	45
A state's right to regulate firearm ownership among its citizens	%	50	50
The federal government's right to regulate state militias	%	46	54
The federal government's right to regulate who can own firearms	%	41	59
A private citizen's right to form a militia	%	36	64

Note: Percentages may not add up to 100% due to rounding.

SOURCE: "Table 1a. Perceived As Supported/Not Supported by Second Amendment," in *77% of Americans Feel Firearm Rights Should Come with Some Restrictions; 14% Favor No Limitations*, The Harris Poll, April 1, 2014, http://www.harrisinteractive.com/vault/Harris%20Poll%2030%20-%20Second%20Amendment_4.1.2014.pdf (accessed July 14, 2014)

signed this law to enable the rapid creation of a militia when needed, but the law also permitted carrying arms in self-defense and forbade the use of arms only when the intention was to "terrify the King's subjects." In 1328, under the reign of King Edward III (1312–1377), Parliament enacted the Statute of Northampton, which prohibited the carrying of arms in public places but did not overrule the right to carry arms in self-defense.

EARLY GUN CONTROL LAWS

The 17th century was a period of great turmoil in England, as Parliament and the monarchy struggled for control of the government. A critical issue related to the civil wars that erupted in 1642 was whether the king or Parliament had the right to control the militia. When the wars ended in early 1660, England fell briefly under the control of a military government, which authorized its officers to search for and seize all arms owned by Catholics or any other people deemed dangerous. In late 1660, with the coronation of Charles II (1630–1685), the English monarchy was restored, but the battle between Parliament and the monarchy continued.

The Game Act of 1671, an early example of a gun control law, was enacted to keep the ownership of hunting lands and weaponry in the hands of the wealthy and to restrict hunting and gun ownership among the peasants. People without an annual income of at least 40 to 100 pounds could no longer keep weapons, even for self-defense. In 1689 Queen Mary II (1662–1694) and King William III (1650–1702) were installed as corulers of

England. When the pair took their oaths of office, they were presented with a new Bill of Rights (2008, http://avalon.law.yale.edu/17th_century/england.asp), which outlined the relationship of Parliament and the monarchy to the people. This Bill of Rights included a specific right of "Protestants [to] arms for their defence suitable to their conditions and as allowed by law." It also condemned abuses committed by standing armies (armies maintained by the government on a long-term basis, even while at peace) and declared "that the raising or keeping a standing army within the kingdom in time of peace, unless it be with consent of Parliament, is against law." Furthermore, the Bill of Rights removed the word *guns* from the list of items the poor were forbidden to own by the Game Act of 1671. From this time on, the right to keep and bear arms belonged to all Englishmen, whether rich or poor.

THE AMERICAN MILITIA AND THE RIGHT TO BEAR ARMS

Much of U.S. law is rooted in the system of laws developed in England (called common law) because most American colonists came from England, bringing with them English values, traditions, and legal concepts. Many of the English were familiar with the famous judge Sir William Blackstone (1723–1780), who listed in his *Commentaries* (2000, http://press-pubs.uchicago.edu/founders/documents/v1ch16s5.html) the "right of having and using arms for self-preservation and defense." This right, brought to North America by the English, was exercised by the colonists against the English during the Revolutionary War (1775–1783) and was later incorporated into the U.S. Constitution.

During the mid-18th century an increasing British military presence in the colonies alerted colonists to the danger of a standing army. When British soldiers shot and killed five men on the streets of Boston in 1770, an event that became known as the Boston Massacre, colonists grew further concerned. The Boston Massacre became a milestone on the road to the Revolutionary War. In 1775 the British army encountered the Massachusetts militia at Lexington—famously recalled in "Concord Hymn" by the essayist and poet Ralph Waldo Emerson (1803–1882) as "the shot heard round the world"—and the ensuing seizure of colonial arms and munitions convinced other colonies that a militia was necessary to achieve the "security of a free state."

Because the individual colonies did not have enough money to purchase weapons, each man was required to maintain a firearm so he could report immediately for duty and form a militia. It was taken for granted by the colonists that the right to individually possess and bear arms was inseparable from the right to form a militia—without these privileges, the right to organize a militia would have little meaning. Thomas Jefferson (1743–1826) stated, "No freeman shall be debarred the use of arms (within his own lands or tenements)," and Richard Henry Lee (1732–1794) of Virginia observed that "to preserve liberty, it is essential that the whole body of the people always possess arms."

In *The Federalist, No. 29* ([January 9, 1788] November 10, 2013, http://www.constitution.org/fed/federa29 .htm), one of a series of papers written after the Revolutionary War to convince the colonists to ratify the Constitution, Alexander Hamilton (1755?–1804) spoke of the right to bear arms in the sense of an unorganized militia, which consisted of the "people at large." He suggested that this militia could mobilize against a standing army if the army usurped the government's authority or if it supported a tyrannical government. Such a standing army, declared Hamilton, could "never be formidable to the liberties of the people while there is a large body of citizens, little, if at all, inferior to them in discipline and the use of arms, who stand ready to defend their own rights and those of their fellow-citizens."

James Madison (1751–1836) attributed the colonial victory to armed citizens. In *The Federalist, No. 46* ([January 29, 1788] November 10, 2013, http://www.constitution.org/fed/federa46.htm), he wrote, "Besides the advantage of being armed, which the Americans possess over the people of almost every other nation, the existence of subordinate governments, to which the people are attached, and by which the militia officers are appointed, forms a barrier against the enterprises of ambition, more insurmountable than any which a simple government of any form can admit of. Notwithstanding the military establishments in the several kingdoms of Europe, which are carried as far as the public resources will bear, the governments are afraid to trust the people with arms."

THE U.S. CONSTITUTION, THE RIGHT TO BEAR ARMS, AND THE MILITIA

The Revolutionary War ended in 1783. In 1787, 39 men gathered in Philadelphia, Pennsylvania, to sign the newly written Constitution. Three refused to sign because the document did not include a Bill of Rights. One reluctant signer protested that, without a Bill of Rights, Congress "at their pleasure may arm or disarm all or any part of the freemen of the United States."

By 1791 James Madison had written the 10 amendments to the Constitution that became known as the Bill of Rights. He was influenced by state bills of rights and many amendment suggestions from the state conventions that ratified the Constitution. Overall, four basic beliefs were assimilated into the Second Amendment: the right of the individual to possess arms, the fear of a professional army, the dependence on militias regulated by the individual states, and the control of the military by civilians.

According to the Constitution Society, in *Documents on the First Congress Debate on Arms and Militia* (February 25, 2005, http://www.constitution.org/mil/militia _debate_1789.htm), the ratifying state conventions offered similar suggestions about the militia and the right to bear arms. Although New Hampshire did not mention the militia, it did state that "no standing Army shall be Kept up in time of Peace unless with the consent of three fourths of the Members of each branch of Congress, nor shall Soldiers in Time of Peace be Quartered upon private Houses without the consent of the Owners."

Another New Hampshire amendment read, "Congress shall never disarm any Citizen unless such as are or have been in Actual Rebellion." Maryland proposed five separate amendments, which Virginia consolidated by stating: "That the people have a right to keep and bear arms; that a well regulated Militia composed of the body of the people trained to arms is the proper, natural and safe defence of a free State. That standing armies in time of peace are dangerous to liberty, and therefore ought to be avoided, as far as the circumstances and protection of the Community will admit; and that in all cases the military should be under strict subordination to and governed by the Civil power."

The New York convention offered more than 50 amendments, including the following: "That the People have a right to keep and bear Arms; that a well regulated Militia, including the body of the People capable of bearing Arms, is the proper, natural and safe defence of a free State."

As early as 1776 Congress had advised the colonies to form new governments "such as shall best conduce to the happiness and safety of their constituents." Within a year after the Declaration of Independence was signed, nearly every state had drawn up a new constitution. The constitutions of several states guaranteed the rights of individuals to bear arms but forbade the maintenance of a standing army. In *State Constitutional Right to Keep and Bear Arms Provisions, by Date* (2006, http://www.law .ucla.edu/volokh/beararms/statedat.htm), Eugene Volokh of the University of California, Los Angeles, notes that Pennsylvania's constitution ensured "that the people have a right to bear arms, for the defence of the State; and, as standing armies, in time of peace, are dangerous to liberty, they ought not to be kept up; and that the military should be kept under strict subordination to, and governed by, the civil power."

Because of their fear of tyranny and repression by a standing army, the colonists preferred state militias to provide protection and order. Such militias could also act as counterbalances against any national standing army. Some people believe that the individual right to bear arms was guaranteed by state laws providing for a militia made up of people trained to use arms.

The Bill of Rights was adopted in 1791. The Second Amendment states: "A well regulated Militia, being necessary to the security of a free State, the right of the people to keep and bear Arms, shall not be infringed."

THE MODERN DEBATE

Efforts to control gun ownership are usually a response to gun-related violence, and most of these initiatives focus on weapons that are perceived to be designed specifically to kill other humans, rather than on guns designed for hunting. Table 1.2 lists the types of firearms

TABLE 1.2

Types of firearms used in crime

Types	
Handgun	A weapon designed to fire a small projectile from one or more barrels when held in one hand with a short stock designed to be gripped by one hand.
Revolver	A handgun that contains its ammunition in a revolving cylinder that typically holds five to nine cartridges, each within a separate chamber. Before a revolver fires, the cylinder rotates, and the next chamber is aligned with the barrel.
Pistol	Any handgun that does not contain its ammunition in a revolving cylinder. Pistols can be manually operated or semiautomatic. A semiautomatic pistol generally contains cartridges in a magazine located in the grip of the gun. When the semiautomatic pistol is fired, the spent cartridge that contained the bullet and propellant is ejected, the firing mechanism is cocked, and a new cartridge is chambered.
Derringer	A small single- or multiple-shot handgun other than a revolver or semiautomatic pistol.
Rifle	A weapon intended to be fired from the shoulder that uses the energy of the explosive in a fixed metallic cartridge to fire only a single projectile through a rifled bore for each single pull of the trigger.
Shotgun	A weapon intended to be fired from the shoulder that uses the energy of the explosive in a fixed shotgun shell to fire through a smooth bore either a number of ball shot or a single projectile for each single pull of the trigger.
Firing action	
Fully automatic	Capability to fire a succession of cartridges so long as the trigger is depressed or until the ammunition supply is exhausted. Automatic weapons are considered machine guns subject to the provisions of the National Firearms Act.
Semiautomatic	An autoloading action that will fire only a single shot for each single function of a trigger.
Machine gun	Any weapon that shoots, is designed to shoot, or can be readily restored to shoot automatically more than one shot without manual reloading by a single function of the trigger.
Submachine gun	A simple fully automatic weapon that fires a pistol cartridge that is also referred to as a machine pistol.
Ammunition	
Caliber	The size of the ammunition that a weapon is designed to shoot, as measured by the bullet's approximate diameter in inches in the United States and in millimeters in other countries. In some instances, ammunition is described with additional terms, such as the year of its introduction (.30/06) or the name of the designer (.30 Newton). In some countries, ammunition is also described in terms of the length of the cartridge case (7.62×63 mm).
Gauge	For shotguns, the number of spherical balls of pure lead, each exactly fitting the bore, that equals one pound.

SOURCE: Marianne W. Zawitz, "What Are the Different Types of Firearms?" in *Guns Used in Crime*, U.S. Department of Justice, Office of Justice Programs, Bureau of Justice Statistics, July 1995, http://bjs.ojp.usdoj.gov/content/pub/pdf/GUIC.PDF (accessed July 14, 2014)

that are used in crime and defines many of the types of weapons mentioned in this chapter.

An effort to decrease crime during the early 1930s led to an unsuccessful attempt by President Franklin D. Roosevelt (1882–1945) to pass legislation requiring the registration of handguns. Nonetheless, the National Firearms Act was passed in 1934, which imposed a tax on the manufacture and sale of "Title II weapons" and mandated the registration of those weapons. Title II weapons include machine guns, short-barreled rifles, short-barreled shotguns, destructive devices (such as grenades), and a catchall category that includes novelty devices such as pen guns and cane guns. It did not, however, include handguns.

The next major piece of gun control legislation—the Gun Control Act of 1968—was passed after the assassinations of King and Senator Kennedy. The act comprises the Title I portion of the U.S. federal firearms laws, and the National Firearms Act described earlier comprises the Title II portion. Title I has provisions that include requiring serial numbers on all guns, setting standards for gun dealers, prohibiting mail-order and interstate sales of firearms, prohibiting the importation of guns not used for sporting purposes, and setting penalties for carrying and using firearms in crimes of violence or drug trafficking. It also prohibits certain categories of people, such as convicted felons, drug addicts, illegal aliens, and minors, from buying or possessing firearms. However, there was no efficient national system for carrying out background checks until the Brady Handgun Violence Prevention Act was passed in 1993.

The Brady Handgun Violence Prevention Act (the Brady law) was named for James S. Brady (1940–2014), who was an official in the administration of President Ronald Reagan (1911–2004). Brady was shot during a 1981 assassination attempt on the president. The Brady law imposed a five-day waiting period on handgun purchases and a background check on buyers to determine whether they were illegal aliens or had a history of criminal behavior, mental illness, or drug use. It required state and local law enforcement agencies to carry out the background checks until a national system could be established. These early background checks were never effectively carried out because Congress did not provide the funds, and in 1997 the requirement was found unconstitutional by the U.S. Supreme Court under the 10th Amendment (states' rights). The court stated in *Printz v. United States* (521 U.S. 898) that the federal government had no right to order state and local law enforcement agencies to carry out federal programs.

The five-day waiting period and the background check requirement were eventually replaced by a national database for background checks, which became effective in November 1998. Known as the National Instant Criminal

Background Check System (NICS), this computerized system is managed by the Federal Bureau of Investigation (FBI). It is used to perform background checks on people seeking to buy handguns or long guns from federal firearms licensees. (Commonly referred to as an FFL, a federal firearms license is usually required for anyone selling a firearm.) According to the FBI, in "Total NICS Firearm Background Checks: November 30, 1998–October 31, 2014" (November 2014, http://www.fbi.gov/about-us/cjis/nics/reports/total-nics-firearm-background-checks-nov.-30-1998-october-31-2014), a total of 197.1 million background checks had been conducted between the introduction of the NICS and October 31, 2014.

"Collective Rights" versus "Individual Rights"

The modern debate over gun control is described by Robert J. Spitzer in *The Politics of Gun Control* (2008) as a split between a "collective rights" interpretation and an "individual rights" interpretation of the Second Amendment. Spitzer, a professor of political science at the State University of New York, Cortland, calls the Second Amendment a "touchstone of the gun debate."

Proponents of the collective rights argument favor stricter control of guns. They point to the opening words of the Second Amendment—"A well regulated militia, being necessary to the security of a free state"—as an indication that the amendment was intended to guarantee the right of states to maintain militias. They argue that since colonial times the concept of a citizens' militia has fallen from use, having been replaced by the National Guard. By this view, the general population does not need unfettered access to guns because they are no longer expected to form a militia during times of need.

In contrast, proponents of the individual rights argument hold that individuals have a right to keep and bear arms. Opponents of extensive gun control add that the phrase "right of the people" is used in the Second Amendment, as well as in other amendments in the Bill of Rights, and in each case it refers to a right of individuals.

In November 2001 the U.S. Court of Appeals for the Fifth Circuit held in *United States v. Emerson* (No. 99-10331) that the Second Amendment protects the right of individuals to "privately possess and bear their own firearms." Conversely, in December 2002 the U.S. Court of Appeals for the Ninth Circuit ruled in *Silveira v. Lockyer* (No. 01-15098) that the Second Amendment does not grant Americans a personal right to carry firearms. The court's ruling said the purpose of the Second Amendment was to maintain effective state militias. In December 2003 the U.S. Supreme Court declined to hear a challenge of this ruling, leaving the question of gun ownership rights in limbo. Then in November 2007 the High Court announced that it would hear an appeal involving the constitutionality of a District of Columbia law banning the use or possession

of all handguns. *District of Columbia v. Heller* (554 U.S. 570) was argued in March 2008. In June 2008 the court ruled in a 5–4 decision that the Second Amendment guarantees individuals the right to bear arms. This ruling affected federal jurisdictions only, but in June 2010, with the ruling in *McDonald v. Chicago* (561 U.S. ___ [2010]), the High Court extended the same interpretation of the Second Amendment to include state and local jurisdictions. These rulings represented an enshrining of the individual's right to bear arms to a degree that went beyond any previous rulings since the beginning of the modern gun control era.

Is Gun Control Unconstitutional?

This strengthening of the individual's right to bear arms, however, did not preclude the ability of the federal and state governments to craft and enforce gun control legislation. The majority opinion in *Heller*, written by Justice Antonin Scalia (1936–), noted that "it is not a right to keep and carry any weapon whatsoever in any manner whatsoever and for whatever purpose." For example, the District of Columbia requires that firearms be registered and bans the carrying of guns in public, both openly and concealed. Keeping a loaded gun in one's home is legal. The provision for gun control legislation applies to state and local jurisdictions in the *McDonald* ruling as it does to federal jurisdictions in the *Heller* ruling.

Arguments for and against Gun Control

One of the key issues in the debate over gun control is whether placing greater restrictions on gun ownership will make society safer. Many opponents of extensive gun control think that access to guns makes it possible for law-abiding Americans to protect themselves and deter crime. By contrast, proponents of extensive gun control hold that Americans infrequently use guns for this purpose.

Handguns are a particular point of contention in the gun control debate because they are seen as the weapon of choice for criminals. Advocates of gun rights point out, however, that upstanding citizens use handguns for self-defense and argue that any attempt to control handgun use is unconstitutional. Those opposing this argument reply that no gun control legislation—including legislation affecting handguns—has ever been declared unconstitutional by the Supreme Court under the Second Amendment.

The statistics appear to indicate that millions of Americans use guns to defend themselves annually. Advocates of extensive gun control or gun prohibition argue that the defensive use of handguns does not offset the offensive use of handguns by criminals, which accounts for thousands of deaths and hundreds of thousands of injuries annually. Spitzer, a gun control advocate, notes that "on an individual level, a gun in the hand of a victim can thwart or stop a crime. On an aggregate level, however, more guns mean more gun problems, even though many citizens believe that guns make them safer."

Additionally, Spitzer and other gun control advocates maintain that the high rate of gun ownership in the United States contributes to the suicide rate, noting that guns are more effective and instantaneous than other modes of committing suicide, thus resulting in more deaths and fewer opportunities to intervene. Gun rights advocates tend to argue that suicidal individuals will find other methods for carrying out their plans. Similar disagreement exists on the topic of accidental injuries and fatalities from guns: gun control advocates maintain that these incidents represent preventable injuries and deaths, whereas gun rights advocates point out that many consumer goods result in accidental injuries and deaths but that this does not justify burdening their owners with complicated regulatory systems.

Although many advocates for extensive gun control measures point out that the founding fathers could not have foreseen the existence of the highly effective handguns and assault weapons of the 21st century, the Second Amendment Foundation, a nonprofit group that promotes the right to bear arms, counters that the improved quality of modern weapons does not override the Second Amendment's insistence on the people's right to arm themselves. Likewise, the Second Amendment Foundation and other gun rights groups profess no faith in the ability of stricter gun control measures to keep weapons out of the hands of criminals. They argue that law-abiding citizens' access to guns through legal channels would become limited, whereas criminals would continue to acquire weapons through illegal means.

Spitzer argues that in the interest of national security a compromise must be reached between those who favor gun control and those who favor gun rights. He suggests that citizens should not have access to assault weapons, that access to handguns should be limited, and that ownership of hunting and sporting weapons should be protected. This position is reflected by most gun control advocates, few of whom call for the outright banning of all guns or even of all handguns. Gun rights advocates, meanwhile, often view compromise as a slippery slope that will eventually lead to the wholesale disarming of the U.S. population.

Expert Views

Gun control advocates typically maintain that guns are infrequently used in self-defense and that their widespread use in criminal acts and suicide necessitates stricter regulation. Gun rights advocates typically maintain that guns in the hands of law-abiding citizens make society safer by deterring crime and that further restrictions will only make it harder for these law-abiding citizens to obtain weapons. Experts, like laypeople who participate in the gun control debate, disagree about the degree to which guns are used as protection versus the degree to which they cause unnecessary death and suffering.

Estimates vary widely as to the number of times handguns are used in self-defense annually. Gun rights advocates often point to the research of Gary Kleck, a criminologist at Florida State University, who has maintained since the mid-1990s that guns are used in self-defense roughly 2 million to 2.5 million times per year. Kleck first published these findings in "Armed Resistance to Crime: The Prevalence and Nature of Self-Defense with a Gun" (*Journal of Law and Criminology*, vol. 86, no. 1, 1995), cowritten with Marc Gertz, and he has found his estimates substantiated in numerous successive works, including the book *Armed: New Perspectives on Gun Control* (2001), cowritten with Don B. Kates. If guns really are used in self-defense 2 million to 2.5 million times per year, then claims by gun rights advocates stating that guns prevent more loss of life than they cost would be more than validated. However, Kleck's findings have not been duplicated by other researchers, and they are considered to be extreme overestimates by most others in fields relating to gun use and safety.

David Hemenway of the Harvard Injury Control Research Center (HICRC) is perhaps the most prominent of Kleck's critics, and the HICRC is among the most reputable research centers in the United States associated with the societal effects of firearms. Hemenway has published a number of studies responding to Kleck's early research, including "Survey Research and Self-Defense Gun Use: An Explanation of Extreme Overestimates" (*Journal of Criminal Law and Criminology*, vol. 87, no. 4, Summer 1997), a direct response to the first influential Kleck and Gertz paper. Hemenway maintains that Kleck and Gertz's original research was subject to a "huge overestimation bias" and a failure to validate survey results relative to other statistics.

For example, Hemenway points out that, when checked against FBI statistics regarding the number and circumstances of robberies annually, Kleck and Gertz's findings about defensive gun use in robberies would require one to believe that more than 100% of gun owners subjected to robbery attempts used guns in self-defense, although more than two-thirds of them were asleep at the time of the intruder's arrival and other studies at that time (based on police reports) showed that only around 1.5% of robbery victims used guns in self-defense. Similarly, Kleck and Gertz estimate that people who used guns in self-defense injured approximately 207,000 criminals each year during the early 1990s, although only around 100,000 people, very few of whom were perpetrators of firearm assaults, were treated for firearm injuries in emergency departments nationally.

Far from accepting such critiques, Kleck and Kates counter that studies conducted by Hemenway and others are the ones with major methodological flaws. Kleck and Kates argue that they are the only researchers to have conducted extensive surveys devoted exclusively to the topic of armed self-defense and that their large sample size and unique methodology makes them more reliable than other scholars. They also maintain that arguments such as Hemenway's, which compare specific subsets of crime data (such as robbery statistics) to the Kleck and Kates findings about the number of times guns are used in self-defense, are logically inconsistent. Because the total number of nonfatal gunshot wounds is unknown, Kleck and Kates argue, crime statistics do not capture the number of times crimes are attempted, let alone the number of times guns are used defensively. In their view, comparing their estimates of defensive gun use with incomplete crime statistics is not a sound means of determining whether their estimates are accurate.

Hemenway and other HICRC researchers typically treat firearm ownership and use as a public health issue, an approach maintaining that firearm injuries and deaths represent a threat to the public well-being on the order of major outbreaks of disease, tobacco use, and motor vehicle safety issues—public health issues that have been successfully addressed through vigorous government intervention. The HICRC researchers have conducted wide-ranging reviews of the scholarly literature on firearm ownership rates and violence as well as data analysis for all 50 states and for 26 developed countries, spanning the late 1980s to around 2005.

Key HICRC papers include Lisa M. Hepburn and David Hemenway's "Firearm Availability and Homicide: A Review of the Literature" (*Aggression and Violent Behavior: A Review Journal*, vol. 9, no. 4, July 2004); David Hemenway and Matthew Miller's "Firearm Availability and Homicide Rates across 26 High-Income Countries" (*Journal of Trauma*, vol. 49, vol. 6, December 2000); Matthew Miller, David Hemenway, and Deborah Azrael's "Rates of Household Firearm Ownership and Homicide across US Regions and States, 1988–1997" (*American Journal of Public Health*, vol. 92, no. 12, December 2002); and Miller, Azrael, and Hemenway's "State-Level Homicide

Victimization Rates in the US in Relation to Survey Measures of Household Firearm Ownership, 2001–2003" (*Social Science and Medicine*, vol. 64, no. 3, February 2007). According to the HICRC researchers' reviews of existing literature as well as their numerous data-driven studies, higher rates of gun ownership consistently correlate with higher rates of homicide. This finding holds true in all countries, even when the United States (which has the highest rate of gun ownership among developed countries and is thus a statistical outlier) is excluded; and across U.S. states, it holds true for every age group of residents, even when controlling for poverty and urbanization.

Furthermore, the HICRC group conducts surveys of expert firearm researchers, which are defined as scholars who have been the primary author on at least one peer-reviewed journal article since 2011 on the topic of firearms in the realms of public health, public policy, sociology, or criminology. In the first survey, which was conducted in May 2014 (http://www.hsph.harvard.edu/wp-content/uploads/sites/1264/2014/05/Expert-Survey-1-Results.pdf), the HICRC researchers asked experts to state their level of agreement with the statement "In the United States, having a gun in the home increases the risk of suicide." They also asked the experts to rate the quality of the scientific evidence on this issue. Of 150 experts who completed the survey, 58% strongly agreed with this statement, 26% agreed, 8% neither agreed nor disagreed, 5% disagreed, and 3% strongly disagreed. Concerning the quality of the scientific evidence on this topic, 26% of the researchers said it was very strong, 37% said it was strong, 27% said it was medium, 8% said it was weak, and 2% said it was very weak.

In a second survey, which was conducted in June 2014 (http://www.hsph.harvard.edu/wp-content/uploads/sites/1264/2014/05/Expert-Survey2-Results.pdf), the HICRC researchers asked experts to state their level of agreement with the statement "In the United States, guns are used in self-defense far more often than they are used in crime" and to rate the quality of the scientific evidence on this issue. Of 122 experts who completed the survey, 39% strongly disagreed with this statement, 34% disagreed, 11% neither agreed nor disagreed, 4% agreed, 4% strongly agreed, and 9% professed not to know. The experts' views of the quality of research on this topic were more mixed, however: 12% rated it very strong, 20% strong, 19% medium, 20% weak, 9% very weak, and 19% professed not to know.

CHAPTER 2
HOW MANY GUNS ARE THERE, AND WHO OWNS THEM?

OWNERSHIP BY PRIVATE CITIZENS

There Can Only Be Estimates

"How many guns are there?" is a question that cannot be answered with exact figures for the United States due to differences among the states. Each has its own system of counting and classifying guns. Some states do not require registration of guns, and unregistered guns cannot be included in an official count. In addition, some types of gun data are restricted from public access. The result is that there can only be estimates of the total number of guns that U.S. residents possess.

The ATF Estimates

The Bureau of Alcohol, Tobacco, Firearms, and Explosives (ATF) is a law enforcement organization in charge of reducing violent crime, among other tasks. One of its duties is to keep firearms out of the hands of criminals. The ATF is also responsible for estimating the total number of firearms in the United States. It does this by adding domestic firearms production and imports since 1899, then subtracting firearms exports during the same period. The ATF does not take into account guns that are destroyed or that no longer work. The ATF statistics also do not account for guns that are smuggled into or out of the United States or guns that are manufactured illegally.

According to the ATF, in *Firearms Commerce in the United States—2001/2002* (2002), between 1899 and 1999 an estimated 248 million guns became available for sale in the United States (not including those produced for the military). This number included more than 87 million rifles, 86 million handguns, and 72 million shotguns. The ATF estimates that there were 1.5 million guns produced in the United States in 1950, 3.7 million in 1970, 5.6 million in 1980, 3.8 million in 1990, and 4 million in 1999. The ATF data suggest that the number of imported rifles, shotguns, and handguns combined averaged 1 million per year during the 1990s, with handguns accounting for roughly half of that figure. Exports averaged fewer than 400,000 per year. Putting all estimates together, by the end of 1999 the total number of guns privately owned or available for sale in the United States came to approximately 260 million.

The publication of ATF data on gun sales was restricted by law between 2003 and 2008, as a result of an amendment authored by Representative Todd Tiahrt (1951–; R-KS) and attached to funding bills for the U.S. Departments of Commerce, Justice, and State. The Tiahrt Amendment also made it illegal for the agency to report on firearms tracing statistics (gun trace statistics). The Tiahrt Amendment restrictions were relaxed in 2008 and again in 2010, allowing for increased data sharing among law enforcement agencies and for the public release of some data that had been suppressed. However, provisions preventing the public disclosure of certain gun trace data, as well as data use in lawsuits against the firearms industry, remained in place as of November 2014.

With the resumption of regular ATF reporting on gun sales, it is possible to add to the cumulative gun totals from 1899 to 1999 to arrive at a rough estimate of guns either privately owned or available for sale in the United States as of 2012. Table 2.1, Table 2.2, and Table 2.3 provide statistics from the ATF on the number of firearms manufactured, imported, and exported since 1986. Using yearly totals from these tables, it is possible to determine that 58 million firearms were manufactured in the United States between 2000 and 2012 and that 2.8 million of these were exported to other countries, for a total of 55.2 million guns that were both manufactured and offered for sale in the United States. Additionally, 31.9 million firearms were imported from other countries to the United States, bringing the total number of firearms either privately owned or available for sale between 2000 and 2012 to 87.1 million. This figure can be added

TABLE 2.1

Firearms manufactured, 1986–2012

Calendar year	Pistols	Revolvers	Rifles	Shotguns	Misc. firearms*	Total firearms
1986	662,973	761,414	970,507	641,482	4,558	3,040,934
1987	964,561	722,512	1,007,661	857,949	6,980	3,559,663
1988	1,101,011	754,744	1,144,707	928,070	35,345	3,963,877
1989	1,404,753	628,573	1,407,400	935,541	42,126	4,418,393
1990	1,371,427	470,495	1,211,664	848,948	57,434	3,959,968
1991	1,378,252	456,966	883,482	828,426	15,980	3,563,106
1992	1,669,537	469,413	1,001,833	1,018,204	16,849	4,175,836
1993	2,093,362	562,292	1,173,694	1,144,940	81,349	5,055,637
1994	2,004,298	586,450	1,316,607	1,254,926	10,936	5,173,217
1995	1,195,284	527,664	1,411,120	1,173,645	8,629	4,316,342
1996	987,528	498,944	1,424,315	925,732	17,920	3,854,439
1997	1,036,077	370,428	1,251,341	915,978	19,680	3,593,504
1998	960,365	324,390	1,535,690	868,639	24,506	3,713,590
1999	995,446	335,784	1,569,685	1,106,995	39,837	4,047,747
2000	962,901	318,960	1,583,042	898,442	30,196	3,793,541
2001	626,836	320,143	1,284,554	679,813	21,309	2,932,655
2002	741,514	347,070	1,515,286	741,325	21,700	3,366,895
2003	811,660	309,364	1,430,324	726,078	30,978	3,308,404
2004	728,511	294,099	1,325,138	731,769	19,508	3,099,025
2005	803,425	274,205	1,431,372	709,313	23,179	3,241,494
2006	1,021,260	385,069	1,496,505	714,618	35,872	3,653,324
2007	1,219,664	391,334	1,610,923	645,231	55,461	3,922,613
2008	1,609,381	431,753	1,734,536	630,710	92,564	4,498,944
2009	1,868,258	547,195	2,248,851	752,699	138,815	5,555,818
2010	2,258,450	558,927	1,830,556	743,378	67,929	5,459,240
2011	2,598,133	572,857	2,318,088	862,401	190,407	6,541,886
2012	3,487,883	667,357	3,168,206	949,010	306,154	8,578,610

*Miscellaneous firearms are any firearms not specifically categorized in any of the firearms categories defined on the ATF Form 5300.11 Annual Firearms Manufacturing and Exportation Report (AFMER). (Examples of miscellaneous firearms would include pistol grip firearms, starter guns, and firearm frames and receivers.)
The AFMER report excludes production for the U.S. military but includes firearms purchased by domestic law enforcement agencies. The report also includes firearms manufactured for export.
AFMER data is not published until one year after the close of the calendar year reporting period because the proprietary data furnished by filers is protected from immediate disclosure by the Trade Secrets Act. For example, calendar year 2012 data was due to ATF by April 1, 2013, but not published until January 2014.
ATF = Bureau of Alcohol, Tobacco, and Firearms.

SOURCE: "Exhibit 1. Firearms Manufactured (1986–2012)," in *Firearms Commerce in the United States: Annual Statistical Update, 2014*, U.S. Department of Justice, Bureau of Alcohol, Tobacco, Firearms and Explosives, April 24, 2014, http://www.atf.gov/sites/default/files/assets/statistics/CommerceReport/firearms_commerce_annual_statistical_report_2014.pdf (accessed July 14, 2014)

to the 260 million guns estimated to be in the United States as of 1999, for a total of 347.1 million guns privately owned or available for sale in the United States by the end of 2012.

U.S. FIREARMS MANUFACTURING

Table 2.1 shows the numbers of pistols, revolvers, rifles, and shotguns that were manufactured in the United States between 1986 and 2012. The number of pistols manufactured annually rose from 662,973 in 1986 to 2,093,362 in 1993, an increase of 216%, and then dropped to a low of 626,836 in 2001. Thereafter, pistol production steadily increased, rising to an all-time high of 3,487,883 pistols in 2012.

Rifle manufacture ranged between roughly 1 million and 1.5 million annually between 1986 and 2006, before increasing to a high of 2,248,851 in 2009. (See Table 2.1.) Rifle production fell to 1,830,556 weapons in 2010, a 19% decline from the previous year, before returning to record levels in 2011 (2,318,088) and then dramatically exceeding those levels in 2012 (3,168,206). Meanwhile,

the manufacture of shotguns increased from 641,482 weapons in 1986 to a high of 1,254,926 in 1994, before falling to a low of 630,710 in 2008. Shotgun manufacturing rebounded in the years that followed, reaching 949,010 in 2012, while remaining below record levels. The number of revolvers manufactured declined 40% between 1986 and 1991, from 761,414 to 456,966, and then leveled off in the 300,000 to 400,000 range between 1997 and 2007. Revolver manufacturing rebounded beginning in 2009 (547,195) and continued to increase in the following years, reaching 667,357 in 2012.

Excluded from the totals detailed in Table 2.1 are National Firearms Act (NFA) weapons. NFA weapons include machine guns, short-barreled rifles, short-barreled shotguns, silencers, and destructive devices. (See Chapter 1 for a description of the National Firearms Act of 1934. In addition, Table 1.2 in Chapter 1 lists and describes several types of firearms.) In 1986 Congress banned the manufacture of machine guns for private sale, but they can be manufactured for export, for the military, and for law enforcement personnel. Nearly 3.7 million NFA weapons were registered in the United States in

TABLE 2.2

Firearms imported, 1986–2013

Calendar year	Shotguns	Rifles	Handguns	Total
1986	201,000	269,000	231,000	701,000
1987	307,620	413,780	342,113	1,063,513
1988	372,008	282,640	621,620	1,276,268
1989	274,497	293,152	440,132	1,007,781
1990	191,787	203,505	448,517	843,809
1991	116,141	311,285	293,231	720,657
1992	441,933	1,423,189	981,588	2,846,710
1993	246,114	1,592,522	1,204,685	3,043,321
1994	117,866	847,868	915,168	1,880,902
1995	136,126	261,185	706,093	1,103,404
1996	128,456	262,568	490,554	881,578
1997	106,296	358,937	474,182	939,415
1998	219,387	248,742	531,681	999,810
1999	385,556	198,191	308,052	891,799
2000	331,985	298,894	465,903	1,096,782
2001	428,330	227,608	710,958	1,366,896
2002	379,755	507,637	741,845	1,629,237
2003	407,402	428,837	630,263	1,466,502
2004	507,050	564,953	838,856	1,910,859
2005	546,403	682,100	878,172	2,106,675
2006	606,820	659,393	1,166,309	2,432,522
2007	725,752	631,781	1,386,460	2,743,993
2008	535,960	602,364	1,468,062	2,606,386
2009	558,679	864,010	2,184,417	3,607,106
2010	509,913	547,449	1,782,585	2,839,947
2011	529,056	998,072	1,725,276	3,252,404
2012	973,465	1,243,924	2,627,201	4,844,590
2013	936,235	1,507,776	3,095,528	5,539,539

ATF = Bureau of Alcohol, Tobacco, and Firearms

Note: Statistics prior to 1992 are for fiscal years; 1992 is a transition year with five quarters.

SOURCE: "Exhibit 3. Firearms Imports (1986–2013)," in *Firearms Commerce in the United States: Annual Statistical Update, 2014*, U.S. Department of Justice, Bureau of Alcohol, Tobacco, Firearms and Explosives, April 24, 2014, http://www.atf.gov/sites/default/files/assets/statistics/CommerceReport/firearms_commerce_annual_statistical_report_2014.pdf (accessed July 14, 2014)

2014, including 512,790 machine guns. (See Table 2.4.) Texas (337,309), California (292,877), and Virginia (248,939) had the most NFA weapons registered in 2014.

FIREARMS IMPORTS

The United States imports a sizeable number of guns each year. According to the Gun Control Act of 1968, imported firearms must be "generally recognized as particularly suitable for or readily adaptable to sporting purposes, excluding surplus military firearms." Gun import statistics from the U.S. International Trade Commission show that firearm imports for the U.S. civilian market more than quadrupled between 1986 and 1993, from 701,000 to 3,043,321. (See Table 2.2.) Firearms imports, however, declined by more than a million units the following year and remained below 1.5 million annually through 2001. The first decade of the 21st century saw an increase in the number of firearms imported into the United States, surpassing 2 million units in 2005 and reaching an all-time high of 3,607,106 firearms imported in 2009. After falling from this record level, imports

climbed dramatically to 5,539,539 in 2013, more than double the number imported just five years earlier.

Table 2.5 shows the countries from which firearms were imported in 2013. The top three source countries for imported firearms—Brazil (975,489 total firearms), Austria (954,388), and Germany (654,901)—supplied nearly half (47%) of all firearms imported into the United States. Handguns (3,095,528) were the most common type of imported firearm, accounting for 56% of total imports. Austria was the leading supplier of imported handguns (932,117), followed by Germany (518,150), Brazil (452,165), and Croatia (451,657). Brazil (404,234) and Canada (369,512) were the leading source countries for imported rifles, while Turkey (306,312), China (234,486), and Italy (212,557) were the leading source countries for imported shotguns.

FIREARMS EXPORTS

Table 2.2 shows the number of firearms that were exported by U.S. manufacturers between 1986 and 2012. Between 1986 and 1993 the export of firearms nearly doubled from 217,448 to 431,204. From that peak, however, firearm exports began a period of decline, reaching a low of 139,920 in 2004. After 2004 firearms exports increased significantly in 2005 (194,682) and then nearly doubled in 2006 (367,521) but did not sustain that level. The number of firearms exported in 2012 was 287,554.

AUTOMATIC AND SEMIAUTOMATIC FIREARMS

The Firearms Owners' Protection Act, which amended the Gun Control Act of 1968, was signed into law in May 1986 and banned private ownership of any machine gun not already lawfully owned. Machine guns are fully automatic weapons, which means they can fire a steady stream of bullets. (See Table 1.2 in Chapter 1.)

Semiautomatic guns manufactured to look like machine guns, commonly referred to as "assault weapons," are frequently confused with machine guns by the media and the public. The term *assault weapons* generally refers to military-style semiautomatic firearms. These weapons look like guns used in the military, but because they are semiautomatic (able to fire only a single shot for each pull of the trigger), they are not actual military-grade weapons. Nevertheless, the term *assault weapons* is often used by the media and the public to refer to machine guns, which are military weapons.

Although assault weapons do not rise to the lethality level of military-grade machine guns, they do typically enable greater lethality than firearms used for hunting, especially when they are outfitted with accessories such as large magazines (cartridge holders that feed bullets automatically into a gun's chamber). Many mass shooters have used assault weapons with magazines, enabling

TABLE 2.3

Firearms exported, 1986–2012

Calendar year	Pistols	Revolvers	Rifles	Shotguns	Misc. firearms*	Total firearms
1986	16,511	104,571	37,224	58,943	199	217,448
1987	24,941	134,611	42,161	76,337	9,995	288,045
1988	32,570	99,289	53,896	68,699	2,728	257,182
1989	41,970	76,494	73,247	67,559	2,012	261,282
1990	73,398	106,820	71,834	104,250	5,323	361,625
1991	79,275	110,058	91,067	117,801	2,964	401,165
1992	76,824	113,178	90,015	119,127	4,647	403,791
1993	59,234	91,460	94,272	171,475	14,763	431,204
1994	93,959	78,935	81,835	146,524	3,220	404,473
1995	97,969	131,634	90,834	101,301	2,483	424,221
1996	64,126	90,068	74,557	97,191	6,055	331,997
1997	44,182	63,656	76,626	86,263	4,354	275,081
1998	29,537	15,788	65,807	89,699	2,513	203,344
1999	34,663	48,616	65,669	67,342	4,028	220,318
2000	28,636	48,130	49,642	35,087	11,132	172,627
2001	32,151	32,662	50,685	46,174	10,939	172,611
2002	22,555	34,187	60,644	31,897	1,473	150,756
2003	16,340	26,524	62,522	29,537	6,989	141,912
2004	14,959	24,122	62,403	31,025	7,411	139,920
2005	19,196	29,271	92,098	46,129	7,988	194,682
2006	144,779	28,120	102,829	57,771	34,022	367,521
2007	45,053	34,662	80,594	26,949	17,524	204,782
2008	54,030	28,205	104,544	41,186	523	228,488
2009	56,402	32,377	61,072	36,455	8,438	194,744
2010	80,041	25,286	76,518	43,361	16,771	241,977
2011	121,035	23,221	79,256	54,878	18,498	296,888
2012	128,313	19,643	81,355	42,858	15,385	287,554

*Miscellaneous firearms are any firearms not specifically categorized in any of the firearms categories defined on the ATF Form 5300.11 Annual Firearms Manufacturing and Exportation Report (AFMER). (Examples of miscellaneous firearms would include pistol grip firearms, starter guns, and firearm frames and receivers.)
Note: The AFMER report excludes production for the U.S. military but includes firearms purchased by domestic law enforcement agencies.
ATF = Bureau of Alcohol, Tobacco, and Firearms.

SOURCE: "Exhibit 2. Firearms Manufacturers' Exports (1986–2012)," in *Firearms Commerce in the United States: Annual Statistical Update, 2014*, U.S. Department of Justice, Bureau of Alcohol, Tobacco, Firearms and Explosives, April 24, 2014, http://www.atf.gov/sites/default/files/assets/statistics/CommerceReport/firearms_commerce_annual_statistical_report_2014.pdf (accessed July 14, 2014)

them to shoot many times without having to pause for reloading. Those in favor of gun control often argue that semiautomatic assault weapons are inappropriately effective killing tools suitable only for criminal purposes and that these weapons should accordingly be banned. Defenders of the right to own assault weapons typically consider assault weapons to be no more deadly than hunting rifles, and they point to the applicability of assault weapons for hunting, informal target shooting, and competitive target shooting.

In 1989 the ATF issued an order banning the importation of 43 models of semiautomatic assault-type guns following a schoolyard shooting in Stockton, California, that killed five people and wounded 29 others. The Violent Crime Control Act of 1994 banned the sale and possession of 19 assault-type firearms and copycat models, including the Uzi, the TEC-9, and the Street Sweeper. The act also limited the capacity of newly manufactured magazines to 10 bullets. (Chapter 3 describes this law in more detail.)

The assault weapons ban expired in September 2004. Bills were introduced in the U.S. House of Representatives in 2007 and 2008 to reinstate and extend the assault weapons ban, but neither bill became law. During the presidential campaigns of 2008 and 2012 the possibility of reinstating the ban on assault weapons was raised. These attempts took on new urgency following the Newtown, Connecticut, mass shooting of December 2012, in which 20 first graders and six school staff members were fatally shot with a semiautomatic assault weapon. However, the Newtown shooting did not substantially change the political feasibility of reinstating the ban. Senator Dianne Feinstein's (1933–; D-CA) Assault Weapons Ban of 2013 won only 40 votes in the U.S. Senate in April of that year, and no further legislation appeared likely to pass either chamber of the legislature in the foreseeable future.

GUNS IN THE HOME
Percentages of Adults Having Guns in the Home

Figure 2.1 shows the trend in adult gun ownership between 1960 and 2013. The percentage of adult Americans answering "yes" to the question "Do you have a gun in your home?" has varied since 1960, although the general trend has been downward. During the 1960s about half of all respondents reported having a gun in their home, and by the early 1980s only 40% answered

TABLE 2.4

National Firearms Act registered weapons, by state, March 2014

State	Any other weapon[a]	Destructive device[b]	Machine gun[c]	Silencer[d]	Short barreled rifle[e]	Short barreled shotgun[f]	Total
Alabama	1,143	62,117	17,463	11,967	1,915	2,139	96,744
Alaska	318	4,164	1,643	2,919	919	1,204	11,167
Arkansas	588	44,522	5,059	9,609	1,712	1,037	62,527
Arizona	1,128	75,858	15,649	21,382	7,162	1,951	123,130
California	3,806	230,410	28,822	8,907	8,069	12,863	292,877
Colorado	943	40,705	6,499	10,535	3,024	1,472	63,178
Connecticut	680	10,650	20,606	6,375	1,463	966	40,740
District of Columbia	69	36,453	4,355	212	760	1,048	42,897
Delaware	33	2,311	597	293	166	507	3,907
Florida	3,366	107,466	31,501	39,613	10,958	6,924	199,828
Georgia	1,787	56,103	27,536	43,958	5,321	10,707	145,412
Hawaii	34	6,461	377	117	55	61	7,105
Iowa	879	12,791	3,405	414	442	949	18,880
Idaho	597	17,696	3,981	14,539	1,778	428	39,019
Illinois	971	86,775	26,014	1,348	1,512	1,675	118,295
Indiana	1,539	38,795	18,377	22,223	3,129	8,637	92,700
Kansas	691	20,839	3,533	4,493	1,506	864	31,926
Kentucky	1,062	23,979	12,204	18,481	1,813	1,701	59,240
Louisiana	534	48,592	6,583	10,088	2,258	1,613	69,668
Massachusetts	843	13,827	6,911	8,486	1,680	935	32,682
Maryland	967	47,153	25,163	9,103	2,448	3,898	88,732
Maine	572	2,775	4,476	1,728	1,528	430	11,509
Michigan	1,092	23,005	11,193	5,627	804	1,134	42,855
Minnesota	2,643	38,553	7,228	719	1,464	1,051	51,658
Missouri	1,341	27,148	8,863	9,199	2,615	2,384	51,550
Mississippi	413	8,189	3,986	6,072	1,002	727	20,389
Montana	439	3,158	2,163	4,571	744	386	11,461
North Carolina	862	77,564	10,708	13,461	4,080	2,713	109,388
North Dakota	196	1,727	1,506	2,834	356	244	6,863
Nebraska	731	5,843	2,131	3,483	895	781	13,864
New Hampshire	429	3,944	7,267	4,568	2,669	407	19,284
New Jersey	422	39,868	7,058	1,016	802	2,504	51,670
New Mexico	297	74,293	3,718	4,124	1,400	639	84,471
Nevada	744	32,917	7,410	9,104	3,411	850	54,436
New York	1,876	38,665	9,333	3,010	3,771	7,698	64,353
Ohio	1,818	74,930	18,762	26,566	4,630	5,284	131,990
Oklahoma	1,147	14,469	8,412	19,317	2,516	1,558	47,419
Oregon	1,515	19,144	6,499	13,304	2,984	1,365	44,811
Pennsylvania	2,063	133,091	17,714	20,629	5,181	12,511	191,189
Rhode Island	41	3,056	630	30	104	112	3,973
South Carolina	677	27,640	7,161	13,952	2,095	3,761	55,286
South Dakota	349	3,573	1,560	3,651	361	183	9,677
Tennessee	1,566	36,179	12,840	13,349	3,598	5,787	73,320
Texas	6,408	190,192	31,493	86,579	15,533	7,104	337,309
Utah	440	14,804	6,306	12,229	2,484	1,227	37,490
Virginia	2,620	177,233	31,825	21,718	9,069	6,474	248,939
Vermont	225	2,365	1,084	70	182	106	4,032
Washington	1,814	37,889	4,075	14,018	1,437	776	60,009
Wisconsin	754	27,851	6,628	6,347	1,969	1,156	44,705
West Virginia	437	9,563	2,521	3,357	962	589	17,429
Wyoming	300	109,127	1,747	2,040	454	384	114,052
Other US territories	6	320	215	16	11	47	615
Total	**56,215**	**2,246,742**	**512,790**	**571,750**	**137,201**	**131,951**	**3,656,649**

"yes" to the question. The percentage answering "yes" rose through the late 1980s and reached 51% in the early 1990s, before settling in a range between 34% and 43% through the late 1990s and into the second decade of the 21st century. In 2013, 37% of adult Americans reported having a gun in their home.

Table 2.6, which shows the results of the General Social Surveys conducted between 1973 and 2012 by the National Opinion Research Center (NORC) at the University of Chicago, yields a similar picture of decreasing firearm prevalence in American homes. NORC surveys

find that the percentage of adults who had a gun in the home decreased from 47.3% in 1973 to 32.5% in 2000 and then rose to 40.6% in 2012. In spite of the general trend toward decreased firearm prevalence in home, the 2010 and 2012 figures reflect higher levels of gun ownership than at any point since the early 1990s.

Table 2.6 also breaks down survey respondents according to the type of gun. The survey results indicate that, while shotguns and rifles were the most common type of firearms in American households during the 1970s and 1980s, the prevalence of these types of firearms has fallen

ᵃThe term "any other weapon" means any weapon or device capable of being concealed on the person from which a shot can be discharged through the energy of an explosive, a pistol or revolver having a barrel with a smooth bore designed or redesigned to fire a fixed shotgun shell, weapons with combination shotgun and rifle barrels 12 inches or more, less than 18 inches in length, from which only a single discharge can be made from either barrel without manual reloading, and shall include any such weapon which may be readily restored to fire. Such term shall not include a pistol or a revolver having a rifled bore, or rifled bores, or weapons designed, made, or intended to be fired from the shoulder and not capable of firing fixed ammunition.
ᵇDestructive device generally is defined as (a) Any explosive, incendiary, or poison gas (1) bomb, (2) grenade, (3) rocket having a propellant charge of more than 4 ounces, (4) missile having an explosive or incendiary charge of more than one-quarter ounce, (5) mine, or (6) device similar to any of the devices described in the preceding paragraphs of this definition; (b) any type of weapon (other than a shotgun or a shotgun shell which the Director finds is generally recognized as particularly suitable for sporting purposes) by whatever name known which will, or which may be readily converted to, expel a projectile by the action of an explosive or other propellant, and which has any barrel with a bore of more than one-half inch in diameter; and (c) any combination of parts either designed or intended for use in converting any device into any destructive device described in paragraph (a) or (b) of this section and from which a destructive device may be readily assembled. The term shall not include any device which is neither designed nor redesigned for use as a weapon; any device, although originally designed for use as a weapon, which is redesigned for use as a signaling, pyrotechnic, line throwing, safety, or similar device; surplus ordnance sold, loaned, or given by the Secretary of the Army pursuant to the provisions of section 4684(2), 4685, or 4686 of title 10, United States Code; or any other device which the Director finds is not likely to be used as a weapon, is an antique, or is a rifle which the owner intends to use solely for sporting, recreational, or cultural purposes.
ᶜMachine gun is defined as any weapon which shoots, is designed to shoot, or can be readily restored to shoot, automatically more than one shot, without manual reloading, by a single function of the trigger. The term shall also include the frame or receiver of any such weapon, any part designed and intended solely and exclusively, or combination of parts designed and intended, for use in converting a weapon into a machine gun, and any combination of parts from which a machine gun can be assembled if such parts are in the possession or under the control of a person.
ᵈSilencer is defined as any device for silencing, muffling, or diminishing the report of a portable firearm, including any combination of parts, designed or redesigned, and intended for the use in assembling or fabricating a firearm silencer or firearm muffler, and any part intended only for use in such assembly or fabrication.
ᵉShort-barreled rifle is defined as a rifle having one or more barrels less than 16 inches in length, and any weapon made from a rifle, whether by alteration, modification, or otherwise, if such weapon, as modified, has an overall length of less than 26 inches.
ᶠShort-barreled shotgun is defined as a shotgun having one or more barrels less than 18 inches in length, and any weapon made from a shotgun, whether by alteration, modification, or otherwise, if such weapon as modified has an overall length of less than 26 inches.

SOURCE: "Exhibit 8. National Firearms Act Registered Weapons by State (March 2014)," in *Firearms Commerce in the United States: Annual Statistical Update, 2014*, U.S. Department of Justice, Bureau of Alcohol, Tobacco, Firearms and Explosives, April 24, 2014, http://www.atf.gov/sites/default/files/assets/statistics/ CommerceReport/firearms_commerce_annual_statistical_report_2014.pdf (accessed July 14, 2014)

over time, while the prevalence of pistols and revolvers has remained relatively consistent. As of 2012, 21.7% of adults had a pistol or revolver in their home, 24.7% had a shotgun, and 24.3% had a rifle.

Research on Guns in the Home and Safety

Does the presence of guns in the home increase or decrease the overall safety of a household? This question has long been a point of contention between gun rights and gun control advocates. Guns in the home can be used to fend off intruders, and gun rights advocates typically maintain that the increased safety brought about by their use in such circumstances outweighs any negative safety effects they may have. However, guns can also be used in conflicts between household members, be accidentally discharged, and be used to commit suicide. Numerous studies have attempted to quantify the effect on safety of having guns in the home, and most have found that the injury or death of a household member or other innocent person is a much more likely outcome than the successful protection of the home against an intruder.

In "Injuries and Deaths Due to Firearms in the Home" (*Journal of Trauma*, vol. 45, no. 2, August 1998), a widely cited study on the relationship between guns in the home and overall safety, Arthur L. Kellerman et al. report on their analysis of police, medical examiner, emergency medical, and hospital records pertaining to shootings in Memphis, Tennessee; Seattle, Washington; and Galveston, Texas. Of the 626 shootings that occurred in these three cities during the study period, 54 were unintentional shootings, 118 were attempted or completed suicides, and 438 were assaults or homicides. By contrast, only 13 shootings were legally justifiable or motivated by self-defense, and three of those involved law enforcement officers. The researchers accordingly conclude that "guns kept in homes are more likely to be involved in a fatal or nonfatal accidental shooting, criminal assault, or suicide attempt than to be used to injure or kill in self-defense."

Subsequent studies have reached varying quantitative conclusions, but most have been in broad agreement regarding their overall conclusions. Linda L. Dahlberg, Robin M. Ikeda, and Marcie-jo Kresnow of the Centers for Disease Control and Prevention find in "Guns in the Home and Risk of a Violent Death in the Home: Findings from a National Study" (*American Journal of Epidemiology*, vol. 160, no. 10, November 15, 2004) that based on their analysis of government mortality statistics, "Those persons with guns in the home were at greater risk than those without guns in the home of dying from a homicide in the home. ... They were also at greater risk of dying from a firearm homicide, but risk varied by age and whether the person was living with others at the time of death. The risk of dying from a suicide in the home was greater for males in homes with guns than for males without guns in the home.... Persons with guns in the home were also more likely to have died from suicide committed with a firearm than from one committed by using a different method."

In "Guns, Fear, the Constitution, and the Public's Health" (*New England Journal of Medicine*, vol. 358, April 3, 2008), Garen J. Wintemute of the University of

TABLE 2.5

Firearms imported, by country, 2013

	Handguns	Rifles	Shotguns	Total firearms
Brazil	452,165	404,234	119,090	975,489
Austria	932,117	21,653	618	954,388
Germany	518,150	135,381	1,370	654,901
Italy	237,918	53,115	212,557	503,590
Croatia	451,657	0	0	451,657
Turkey	105,757	0	306,312	412,069
Canada	6,030	369,512	5	375,547
China[a]	0	4,155	234,486	238,641
Russia	772	169,112	34,904	204,788
Philippines	140,813	5,909	9,800	156,522
Serbia	50,658	44,672	0	95,330
Argentina	82,635	0	0	82,635
Japan	0	76,399	1,525	77,924
Czech Republic	39,897	26,856	142	66,895
Romania	3,655	44,734	0	48,389
Belgium	14,499	29,920	10	44,429
Finland	0	43,858	0	43,858
Israel	23,979	18,504	0	42,483
Bulgaria	8,397	31,087	0	39,484
Spain	262	17,760	1,620	19,642
United Kingdom	92	4,345	7,204	11,641
Switzerland	5,806	3,890	0	9,696
Poland	8,406	510	0	8,916
Portugal	20	4	6,415	6,439
Ukraine	4,000	0	0	4,000
Korea	3,879	0	0	3,879
Other[b]	1,401	770	177	2,348
Taiwan	0	1,396	0	1,396
United Arab Emirates	1,359	0	0	1,359
Slovak Republic	1,204	0	0	1,204
Totals	**3,095,528**	**1,507,776**	**936,235**	**5,539,539**

[a]On May 26, 1994, the United States instituted a firearms imports embargo against China. Shotguns, however, are exempt from the embargo.
[b]Imports of fewer than 1,000 per country.
Notes: Imports from Afghanistan, Belarus, Burma, China, Cuba, Democratic Republic of Congo, Haiti, Iran, Iraq, Libya, Mongolia, North Korea, Rwanda, Somalia Sudan, Syria, Unita (Angola), Vietnam, may include surplus military curio and relic firearms that were manufactured in these countries prior to becoming proscribed or embargoed and had been outside those proscribed countries for the preceding five years prior to import. Imports may also include those that obtained a waiver from the U.S. State Department. Imports from Georgia, Kazakstan, Kyrgyzstan, Moldova, Russian Federation, Turkmenistan, Ukraine, Uzbekistan are limited to firearms enumerated on the Voluntary Restraint Agreement (VRA).

SOURCE: "Exhibit 5. Firearms Imported into the United States by Country of Manufacture 2013," in *Firearms Commerce in the United States: Annual Statistical Update, 2014*, U.S. Department of Justice, Bureau of Alcohol, Tobacco, Firearms and Explosives, April 24, 2014, http://www.atf.gov/sites/default/files/assets/statistics/CommerceReport/firearms_commerce_annual_statistical_report_2014.pdf (accessed July 14, 2014)

California, Davis, School of Medicine frames the presence of guns in the home as a public health issue and summarizes findings from a variety of researchers to quantify the risks that accompany firearm ownership. He states, "Living in a home where there are guns increases the risk of homicide by 40 to 170% and the risk of suicide by 90 to 460%. Young people who commit suicide with a gun usually use a weapon kept at home, and among women in shelters for victims of domestic violence, two thirds of those who come from homes with guns have had those guns used against them."

According to most researchers, children in homes where guns are present are at particular risk of injury or death. In its policy statement "Firearm-Related Injuries

Affecting the Pediatric Population" (*Pediatrics*, vol. 130, no. 5, November 1, 2012), the American Academy of Pediatrics (AAP) notes that "although rates have declined since the American Academy of Pediatrics (AAP) issued the original policy statement in 1992, firearm-related deaths continue as 1 of the top 3 causes of death in American youth." The AAP goes on to maintain, as part of its official recommendations, that "the most effective measure to prevent suicide, homicide, and unintentional firearm-related injuries to children and adolescents is the absence of guns from homes and communities."

Gary Kleck and Don B. Kates provide a dissenting view in *Armed: New Perspectives on Gun Control* (2001). The authors argue that estimates of defensive gun use are usually low due to errors in research methodology and that guns are used for defense 2 million to 2.5 million times per year in the United States. Kleck and Kates posit that this estimate has been confirmed many times "by all surveys of comparable technical merit." They also contend that the study of guns and violence in general, and the study of guns and self-defense in particular, are plagued by "advocacy scholarship" and "junk science."

CHARACTERISTICS OF FIREARM OWNERS

In *Why Own a Gun? Protection Is Now Top Reason* (March 12, 2013, http://www.people-press.org/files/legacy-pdf/03-12-13%20Gun%20Ownership%20Release.pdf), the Pew Research Center provides an overview of the characteristics of gun owners in the United States based on surveys conducted in February 2013. Pew finds that 37% of survey respondents reported the presence of a gun in their home, 24% of whom reported personally owning the gun and 13% of whom reported that someone else in the household owned the gun. (See Table 2.7.) Nearly half (45%) of the men surveyed reported a gun in the household, and 37% of them claimed to own the gun personally, compared with 30% and 12% of women, respectively. Respondents aged 50 to 64 years (40%) and 65 years and older (40%) owned a gun at a higher rate than those between the ages of 18 and 29 years (35%) and 30 to 49 years (35%). There appeared to be no conclusive correlation between education levels and household gun ownership: the least educated group of respondents, those whose highest level of education was a high school diploma or less (33%), reported a gun in the home at the same rate as the most educated group, those with a postgraduate degree (33%); and college graduates (40%) and those who had attended but not graduated from college (43%) had the highest rates of household gun ownership. Likewise, there were no dramatic differences in the gun-owning rates of parents (34%) versus nonparents (39%), although women who were parents (24%) were significantly less likely to report having a gun in the home than women who were not parents (33%).

FIGURE 2.1

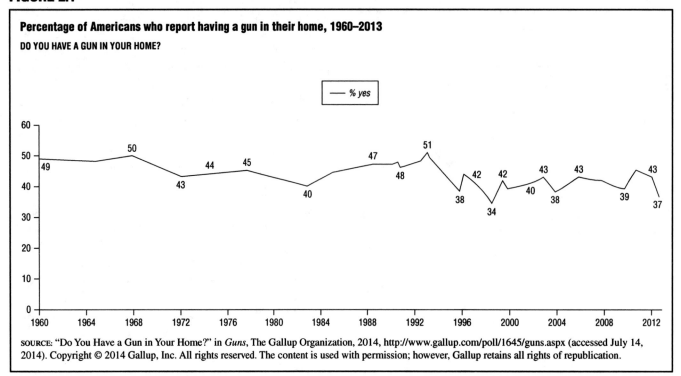

Percentage of Americans who report having a gun in their home, 1960–2013

DO YOU HAVE A GUN IN YOUR HOME?

SOURCE: "Do You Have a Gun in Your Home?" in *Guns*, The Gallup Organization, 2014, http://www.gallup.com/poll/1645/guns.aspx (accessed July 14, 2014). Copyright © 2014 Gallup, Inc. All rights reserved. The content is used with permission; however, Gallup retains all rights of republication.

TABLE 2.6

Percentage of Americans who report having a gun in their home, by type of gun and personal ownership, selected years 1973–2012

	% of adults with gun in home	Of adults with gun in home, % with pistol or revolver	Of adults with gun in home, % with shotgun	Of adults with gun in home, % with rifle	Of adults with gun in home, % who personally own the gun(s)
1973	47.3	19.9	27.6	29.2	—
1976	46.7	21.6	28.1	28.2	—
1980	47.7	23.2	29.6	29.2	61.8
1984	45.2	21.4	27.7	27.3	56.1
1988	40.1	22.6	23.9	24.2	59.9
1990	42.7	23.5	25.9	25.4	66.2
1994	40.7	24.5	24.3	24.5	68.3
1996	40.2	22.3	24.7	23.3	66.2
1998	34.9	19.6	20.8	20.8	65.2
2000	32.5	19.7	18.6	19.7	66.5
2002	33.5	19.5	21.1	19.8	74.9
2004	35.7	20.7	18.9	19.7	70.3
2006	33.2	20.0	19.3	19.1	64.8
2008	34.4	24.3	23.8	23.7	68.3
2010	40.9	21.7	24.9	24.5	64.8
2012	40.6	21.7	24.7	24.3	64.7

SOURCE: Adapted from "Owngun," "Pistol," "Shotgun," "Rifle," and "Rowngun," in *General Social Surveys (GSS) 1972–2012 Cumulative Datafile*, National Opinion Research Center, University of Chicago, 2014, http://sda.berkeley.edu/sdaweb/analysis/?dataset=gss12 (accessed July 14, 2014)

In contrast, the demographic indicators that had the strongest correlations with gun ownership were race and Hispanic origin, U.S. region, and urban/suburban/rural location. (See Table 2.7.) Racial differences in gun ownership rates were pronounced: 46% of white survey respondents reported the presence of a gun in the home, compared with 21% of African American and 17% of Hispanic respondents. Likewise, respondents in the Midwest (45%) and the South (42%) were much more likely to report the presence of a gun in the household than those living in the West (30%) or Northeast (25%). A solid majority of rural respondents (59%) reported having a gun in their home, compared with 28% of urban respondents and 36% of suburban respondents.

Comparing its 2013 survey results with results from a similar survey that was conducted by ABC News and the *Washington Post* in 1999, Pew finds that gun owners' reasons for owning a gun have changed significantly. (See Table 2.8.) In 1999 roughly half (49%) of all gun owners reported owning a gun primarily for the purpose of hunting, and approximately a quarter (26%) gave protection as the primary motivation for gun ownership. By 2013 the proportions of respondents giving these reasons for gun ownership had reversed: 48% said protection was the primary reason they owned a gun and 32% said hunting was the primary reason.

TABLE 2.7

Gun owners by selected demographic characteristics, 2013

	Household owns gun %	Personally own gun %	Someone else
Total	**37**	**24**	**13**
Men	45	37	8
Women	30	12	17
18–29	35	16	19
30–49	35	24	11
50–64	40	30	10
65+	40	29	11
White	46	31	15
Black	21	15	7
Hispanic	17	11	6
College grad+	37	24	12
Post-graduate	33	20	13
College grad	40	27	13
Some college	43	28	15
High school or less	33	22	11
Northeast	25	17	8
Midwest	45	27	18
South	42	29	13
West	30	21	9
Urban	28	18	10
Suburban	36	24	12
Rural	59	39	20
Married	45	31	14
Not married	30	19	11
Parents	34	23	11
Men	45	40	5
Women	24	8	16
Non-parents	39	26	13
Men	45	36	9
Women	33	15	18

SOURCE: "Demographics of Gun Ownership," in *Why Own a Gun? Protection Is Now Top Reason*, Pew Research Center, March 12, 2013, http://www.people-press.org/files/legacy-pdf/03-12-13%20Gun%20Ownership%20Release.pdf (accessed July 24, 2014)

TABLE 2.8

Reasons for owning a gun, 1999 and 2013

What is main reason you own a gun?	1999			2013			2013
	Protection %	Hunting %	Other %	Protection %	Hunting %	Other %	N
All gun owners	26	49	26	48	32	18	421
Men	21	55	24	42	36	20	303
Women	43	26	31	65	21	13	118
18–49	27	47	25	52	29	18	157
50+	22	52	26	45	36	18	260
College grad+	25	49	26	45	28	25	154
Some college	34	45	21	53	31	14	128
High school grad or less	22	51	27	45	37	17	138
Rep/lean Rep	23	50	27	45	39	14	216
Dem/lean Dem	28	49	23	53	28	19	155

SOURCE: "Why Do You Own a Gun?" in *Why Own a Gun? Protection Is Now Top Reason*, Pew Research Center, March 12, 2013, http://www.people-press.org/files/legacy-pdf/03-12-13%20Gun%20Ownership%20Release.pdf (accessed July 24, 2014)

CHAPTER 3
FIREARM LAWS, REGULATIONS, AND ORDINANCES

FEDERAL GOVERNMENT REGULATION OF GUNS

Americans have long debated the issue of federal regulation of firearms. Those in favor of regulation argue that only federal firearm laws can limit access by criminals, juveniles, and other high-risk people, thereby reducing violent crime. Supporters of regulation also contend that without federal laws, states with few firearms restrictions will supply guns illegally to states with stiffer restrictions. Opponents of federal involvement advance Second Amendment arguments against any kind of gun control. They also argue that federal gun laws only cause extra burdens for law-abiding citizens who seek to buy and sell firearms.

Until the 1920s the states made their own decisions about whether and how to regulate firearms. The federal government stepped in following ratification of the 18th Amendment (1919), which made it illegal to manufacture, transport, and sell "intoxicating liquors" in the United States. Known as Prohibition, the ban sparked a national crime wave as gangsters built empires on the illegal manufacture and sale of alcohol. For the first time, crime was seen as a national problem, and people debated solutions at the federal level.

The first federal law regulating firearms was passed in 1927 (18 USC 1715). The law outlawed mailing any firearm other than a shotgun or rifle through the U.S. Postal Service, except for firearms shipped for official law enforcement purposes. This law, which was still in effect as of 2014, was intended to curb the mail-order business in handguns, and it inspired many states to pass their own regulations regarding the sale and use of handguns. Supporters of the federal law tried to get a law passed forbidding the shipment of handguns across state lines by all commercial carriers, but they failed. Instead, commercial carriers other than the U.S. Postal Service were allowed to carry handguns across state lines.

The next federal regulation was the National Firearms Act of 1934, which was designed to make it more difficult to acquire especially dangerous "gangster-type" weapons such as machine guns, sawed-off shotguns, and silencers. The legislation placed heavy taxes on all aspects of the manufacture and distribution of these firearms and required registration of each firearm through the entire production, distribution, and sales process. This law was still in force, with amendments, as of 2014.

Ordinary firearms were the subject of the Federal Firearms Act of 1938. The act required any manufacturer or dealer who sent or received firearms across state lines to have a federal firearms license and to keep a record of the names and addresses of people purchasing firearms. Firearms could not be sent to anyone who was a fugitive from justice or had been convicted of a felony. It became illegal to transport stolen guns from which the manufacturer's mark had been rubbed out or changed. This act was replaced with the passage of the Gun Control Act of 1968.

THE GUN CONTROL ACT OF 1968

The Gun Control Act of 1968 was passed in the wake of the assassinations of President John F. Kennedy (1917–1963), the civil rights leader Martin Luther King Jr. (1929–1968), and Senator Robert F. Kennedy (1925–1968; D-NY). It repealed the Federal Firearms Act of 1938 and amended the National Firearms Act of 1934 by adding bombs and other destructive devices to machine guns and sawed-off shotguns as items strictly controlled by the government. Its purpose was to assist federal, state, and local law enforcement agencies in reducing crime and violence.

The Gun Control Act of 1968 had two major sections. Title I required anyone dealing in firearms or ammunition—whether locally or across state lines—to be federally licensed under tough new standards and to keep records of all commercial gun sales. Title I also prohibited the interstate mail-order sale of all firearms and ammunition, the interstate sale of handguns generally, and the interstate sale of long guns (rifles and shotguns), except under certain conditions. It forbade sales to minors and those with criminal records, outlawed the importation of nonsporting firearms, and established special penalties for the use or carrying of a firearm while committing a crime of violence or drug trafficking. However, Title I did not forbid the importation of unassembled weapons parts, and some individuals and companies were suspected of importing separate firearms parts and then reassembling them into a complete weapon as a means of getting around the law.

Title II, the National Firearms Act, reenacted the 1934 National Firearms Act and extended the Gun Control Act to cover private ownership of destructive devices such as submachine guns, bombs, and grenades. Enforcement of the federal laws became the responsibility of the U.S. Department of the Treasury, which created the Bureau of Alcohol, Tobacco, and Firearms (ATF) in 1972. The ATF has since come under the control of the U.S. Department of Justice and has been renamed the Bureau of Alcohol, Tobacco, Firearms, and Explosives.

Efforts to Amend the Gun Control Act of 1968

After the passage of the Gun Control Act of 1968, Congress found itself under siege from both pro-gun and antigun groups. Firearms owners and dealers complained about ATF enforcement efforts, which seemed to target law-abiding citizens while neglecting criminals with firearms and people selling firearms illegally. Others voiced concern that the act penalized sportsmen. In contrast, those urging tighter control on guns believed the act did not go far enough in keeping firearms out of the hands of criminals. Every Congress from 1968 to 1986 introduced dozens of pieces of legislation to strengthen, repeal, or diminish the requirements of the 1968 act. In 1986 the gun control advocates were successful.

THE FIREARMS OWNERS' PROTECTION ACT OF 1986

In 1986, nearly 20 years after the passage of the Gun Control Act of 1968, Congress passed major legislation that amended the 1968 law. The Firearms Owners' Protection Act of 1986, commonly referred to as the Gun Control Act of 1986, was still in effect as of 2014. The battle over this piece of legislation was fierce. David T. Hardy describes in "The Firearm Owners' Protection Act of 1986: A Historical and Legal Perspective" (*Cumberland Law Review*, vol. 17, 1986) the reactions of those in favor and those opposed to its passage. Those in favor of the act called its passage "necessary to restore fundamental fairness and clarity to our nation's firearms laws." Those opposed called it an "almost monstrous idea" and

a "national disgrace." The following sections compare the 1968 act with the current standards under the 1986 act.

Changes Implemented with the Gun Control Act of 1986 and Further Amendments

PROHIBITED PEOPLE. The 1968 legislation identified the categories of people to whom firearms could not be sold by a federally licensed firearms dealer, also called a federal firearms licensee (FFL). These included convicted felons, drug abusers, and the mentally ill. The 1986 legislation clarified some inconsistencies in the 1968 act and made it unlawful for anyone, whether licensed or not, to sell a gun to a high-risk individual. The definition of the term *high risk* has been refined over the years and includes groups such as felons, fugitives, illegal aliens, and those subject to a restraining order.

The 1996 enactment of the Domestic Violence Offender Gun Ban (generally known as the Lautenberg Amendment after its sponsor Senator Frank R. Lautenberg [1924–2013; D-NJ]) to the Gun Control Act of 1968 expanded the group of people prohibited from legally obtaining a gun. The Lautenberg Amendment made it a felony for anyone convicted of a misdemeanor crime of domestic violence (e.g., assault or attempted assault on a family member) to ship, transport, possess, or receive firearms or ammunition. There was no exception for military personnel or law enforcement officers engaged in official duties unless their record has been expunged (erased). The amendment also made it a felony for anyone to sell or issue a firearm or ammunition to a person with such a conviction. The Lautenberg Amendment was ruled unconstitutional in 1999, but that ruling was reversed two years later. It was challenged and upheld again in 2009 in *United States v. Hayes* (555 U.S. 415), and, as of 2014, the Lautenberg Amendment was still in force.

The Omnibus Consolidated and Emergency Supplemental Appropriations Act of 1999 also amended the Gun Control Act of 1968 by prohibiting aliens admitted under a nonimmigrant visa from obtaining firearms. This included people traveling temporarily in the United States, those studying in the United States who maintained a residence abroad, and some foreign workers. Exceptions included people entering the United States for lawful hunting or sporting purposes, official representatives of a foreign government, and foreign law enforcement officers from friendly foreign governments who entered the United States on official business. In addition, the U.S. attorney general had the authority to waive the prohibition on submission of a petition by an alien.

PURCHASING FIREARMS AND AMMUNITION. The Gun Control Act of 1968 stated that firearms and ammunition could be purchased only on the premises of an FFL and that FFLs were required to record all firearms and ammunition transactions. The 1986 law made the purchase of ammunition and gun components by mail legal. Under the new law firearms dealers were required to keep records only of armor-piercing ammunition sales. Establishments selling only ammunition (no firearms) would not have to be licensed, thus allowing stores that previously may not have carried ammunition because of licensing requirements to do so.

The Treasury and General Government Appropriations Act of 2000 amended the Gun Control Act to require that firearms owners who pawn their weapons must undergo a background check when they seek to redeem the weapons.

CARRYING A GUN BETWEEN JURISDICTIONS. The 1968 act did not address the effects of state and local regulations concerning the intrastate (within a state) transportation of weapons. The 1986 act made it legal to transport any legally owned gun through a jurisdiction where it would otherwise be illegal, provided the possession and transporting of the weapon were legal at the point of origin and the point of destination. In addition, the gun had to be unloaded and placed in a locked container or in the trunk of a vehicle during transport.

FEDERAL CRIMES. The 1968 legislation prohibited the carrying or use of a firearm during or in relation to a federal crime of violence. In addition, anyone convicted of such a crime would be subject to a minimum penalty over and above any sentence received for the primary offense. The 1986 act added serious drug offenses to the category of prohibited crimes, and it doubled the existing penalty for use of a machine gun or a gun equipped with a silencer.

FORFEITURE OF FIREARMS AND AMMUNITION. Before 1986 any firearm or ammunition involved in, used in, or intended to be used in violation of the Gun Control Act or other federal criminal law could be taken away from the gun owner. However, under the Gun Control Act of 1986, forfeiture is no longer automatic. For some offenses, a willful element must be demonstrated; for others, knowledge of an offense is enough. In the case of a firearm being "intended for use" in a violation, "clear and convincing evidence" of the intent must be shown. In addition, only specified crimes now justify forfeiture, including crimes of violence, drug-related offenses, and certain violations of the Gun Control Act. In all cases forfeiture proceedings must begin within 120 days of seizure, and the court will award attorney fees to the owner if the owner wins the case.

CRIMINAL PENALTIES. The 1968 law stated that a demonstration of willfulness was not needed as an element of proof of violation of any provision of the act, whereas the Gun Control Act of 1986 required proof of willful violations and/or knowing violations to prosecute.

The 1986 act also reduced licensee record-keeping violations from felonies to misdemeanors.

LEGAL DISABILITIES. Under the 1968 law, any person who had been convicted of a crime and sentenced to prison for more than one year was restricted from shipping, transporting, or receiving a firearm and could not be granted an FFL. State pardons could not erase the conviction for federal purposes. However, a convicted felon could make a special request to the U.S. secretary of the treasury to be allowed to possess a firearm. The secretary of the treasury had to certify that the possession of a firearm by the convicted felon was not contrary to public interest and safety and that the applicant did not commit a crime involving a firearm or violate federal gun control laws.

Under the 1986 act, however, state pardons can erase convictions for federal purposes, unless the person is specifically denied the right to possess or receive firearms. The act allows those who violated the law to appeal—even those whose crime involved the use of a firearm or the violation of federal gun control laws.

MACHINE GUN FREEZE. The National Firearms Act of 1934 imposed production and transfer taxes and registration requirements on firearms typically associated with criminal activity. These restrictions applied specifically to machine guns; destructive devices such as bombs, missiles, and grenades; and firearm silencers. The definition of a machine gun included "any combination of parts designed and intended for use in converting a weapon to a machine gun."

The Firearms Owners' Protection Act of 1986 made it "unlawful for any person to transfer or possess a machine gun" unless it was manufactured and legally owned before May 19, 1986. In addition, the definition was revised to include any combination of parts "designed and intended solely and exclusively for use in conversion." By refusing to review a lower court's decision in *Farmer v. Higgins* (907 F.2d 1041 [11th Cir. 1990]), the U.S. Supreme Court upheld the ban on machine-gun ownership.

THE LAW ENFORCEMENT OFFICERS' PROTECTION ACT OF 1986

The Law Enforcement Officers' Protection Act of 1986 also amended the 1968 Gun Control Act. This act banned the manufacture or importation of certain varieties of armor-piercing ammunition. Called "cop-killers," these bullets are capable of penetrating police officers' bulletproof vests. The law defined the banned ammunition as handgun bullets made of specific hard metals: tungsten alloys, steel, brass, bronze, iron, beryllium copper, or depleted uranium. (Standard ammunition is made from lead.) It also decreed that the licenses of dealers who knowingly sold such ammunition should be revoked.

In 1994 the Violent Crime Control and Law Enforcement Act broadened the ban to include other metal-alloy ammunition. Both laws limit the sale of such bullets to the U.S. military or to the police.

The legislation banning armor-piercing ammunition was politically significant because it polarized two traditionally allied groups: the National Rifle Association of America (NRA) and the police. The NRA had long assisted in training police officers in marksmanship. The police saw the ammunition ban as a personal issue, because the cop-killer bullets were intended to harm them. A police lobby, the Law Enforcement Steering Committee (LESC), was formed to get this and other legislation passed. The NRA opposed the legislation as originally written because, it said, the legislation would have also banned most of the types of ammunition used for hunting and target shooting. The final version of the legislation exempted bullets made for rifles and sporting purposes.

THE UNDETECTABLE FIREARMS ACT OF 1988

Highly publicized aircraft hijackings during the 1980s prompted Congress to begin hearings in 1986 to determine if plastic firearms represented a danger to airline passengers if used by terrorists. The passage of the Undetectable Firearms Act of 1988 followed. It banned the manufacture, import, sale, transfer, or possession of a plastic firearm. The act also stated: "If the major parts of the firearms do not permit an accurate X-ray picture of the gun's shape, the firearm is [also defined as] a plastic firearm, even if the firearm contains more than 3.7 ounces of electromagnetically detectable metal."

The NRA called the proposed legislation unnecessary and the first step toward a total ban on handguns. The lobbying efforts of police officers on the LESC convinced the U.S. attorney general Edwin Meese (1931–) that plastic weapons posed an unacceptable hazard to public safety and that this bill was a necessary piece of legislation. The ban was renewed and made permanent in December 2003.

THE "TOY GUN LAW" OF 1988

A law requiring that a "toy, look-alike, or imitation firearm shall have as an integral part, permanently affixed, a blaze orange plug inserted in the barrel of such toy, look-alike, or imitation firearm" was passed as section four of the Federal Energy Management Improvement Act of 1988. The law provided for alternative markings if the orange plug could not be used. However, the plug could be removed and the markings painted over.

Ending Sales of Look-Alike Toy Guns

By November 1994 three major toy retailers announced that they would stop selling toy guns designed to look like real guns because the look-alikes had led to tragic consequences. On September 27, 1994, 13-year-old Nicholas Heyward Jr. was shot and killed by a police officer in Brooklyn, New York, when the officer confronted the boy in a dimly lit stairwell. The appearance of the gun and clicking sounds that occurred led the officer to believe that he was about to be shot. He fired and killed the boy. Later that evening in Brooklyn, 16-year-old Jamiel Johnson was seriously wounded by a plainclothes police officer when the boy pointed his look-alike toy pistol at the officer. Kay-Bee Toy Stores, Toys "R" Us, and Bradlees decided to stop selling these guns, even though the sale of them had generated almost $250 million in 1993. A report by the New York Police Department showed that realistic-looking toy guns had been used in 534 felonies up until October 1994.

Despite pledges by toy retailers to halt their sales of realistic-looking toy guns, the sales did not stop. In 2004, after a New York City council investigation found that 20% of all toy and discount stores in the city were still selling the realistic-looking toy guns, Mayor Michael Bloomberg (1942–) announced a renewed effort to eliminate the guns from local stores' shelves. In 2005 the retail stores CVS and Kmart were levied large fines for selling realistic-looking toy guns. In April 2008 the chain store Party City was accused by the New York City Department of Consumer Affairs of having hundreds of realistic-looking toy guns on its store shelves. In October 2008 the chain store reached a $500,000 settlement with the department for violating the city law.

In "New York City Takes Aim at Illegal Sales of Realistic-Looking Toy Guns" (NYDailyNews.com, December 9, 2009), Frank Lombardi reports on a New York City advertising campaign that was conducted between December 2009 and January 2010. The ads were the continuation of New York City's attack on the illegal sales of realistic-looking toy guns and also the result of criminals painting real guns the luminous, bright colors required of toy guns. In the ad the real gun was luminous red and the realistic-looking toy gun was black. The ad was designed to show how difficult it was to determine which gun was the real, deadly weapon. The city also passed legislation to raise its fines for illegal sales of realistic-looking toy guns. Dan Mangan reports in "Fake-Gun Fine Unreal" (NYPost.com, January 17, 2012) that the city demonstrated its commitment to ending sales of realistic-looking toy guns in 2012, when it fined the owner of a Brooklyn discount store $30,000 for stocking six toy sets that included imitation plastic revolvers.

THE GUN-FREE SCHOOL ZONES ACT OF 1990

Congressional passage of the Gun-Free School Zones Act, which was part of the Crime Control Act of 1990, stipulated that it was unlawful for anyone to knowingly possess firearms in school zones. Considered quite strict, the law made it illegal to carry even unloaded firearms in an unlocked case or bag while on public sidewalks in designated school zones. This applied to the sidewalk in front of the firearm owner's residence as well, if that portion of the walkway was within 1,000 feet (305 m) of the grounds of any public or private school—whether or not school was in session. Gun control advocates thought the legislation was needed to send a message to teachers and law enforcement personnel that the federal government was behind them in the effort to get guns out of schools.

Gun rights advocates condemned this attempt to restrict their rights. The act was challenged as unconstitutional and was struck down by the Supreme Court in *United States v. Lopez* (514 U.S. 549 [1995]). The court upheld state and local authority in the regulation of schools and ruled that Congress had exceeded its power in passing the original act. In response to the court's ruling, Congress approved a slightly revised version of the Gun-Free School Zones Act in 1996. The focus of the act was changed from possessing a firearm in a school zone to possessing a firearm "that has moved in or that otherwise affects interstate or foreign commerce" in a school zone. The amended law has since been upheld numerous times in U.S. circuit courts.

THE BRADY HANDGUN VIOLENCE PREVENTION ACT OF 1993

In 1981 a mentally disturbed young man named John W. Hinckley Jr. (1955–)—armed with a handgun—shot and seriously wounded the presidential press secretary James S. Brady (1940–2014) during an assassination attempt on President Ronald Reagan (1911–2004). After the attempt, Hinckley was quickly apprehended, placed in custody, tried, and found not guilty by reason of insanity. His violent act was the impetus behind the Brady Handgun Violence Prevention Act, which was passed by Congress in November 1993 after years of debate.

"Interim Brady": The Five-Day Waiting Period

The Brady Handgun Violence Prevention Act (commonly referred to as the Brady law), which went into effect on February 28, 1994, established a national five-day waiting period for a person to purchase a handgun. This waiting period allowed time for local law enforcement officers to conduct background checks on handgun purchasers for mental instability or criminal records. The waiting time also provided a cooling-off period to reduce gun-related crimes of passion. Provisions for a waiting period and for background checks by local law enforcement officials were interim measures of the Brady law that could be dropped after a computerized national instant criminal identification system was developed. Such a system became operational in the United States in 1998.

Before 1998, within the first day of the waiting period, licensed firearms dealers were required to send a copy of the purchaser's sworn statement to the chief law enforcement officer where the purchaser resided. The purchaser was asked many questions, including whether he or she was a fugitive from justice, addicted to drugs, illegally in the United States, or convicted of a crime. The dealer could complete the sale after five business days unless the police notified the dealer that the sale would violate the law. It also required the police to respond within 20 business days to any request for a written explanation of a request denial. If the sale was not denied, local law enforcement officials were required to destroy the sworn statement and any other record of the transaction within 20 business days.

In 1997 the Brady law's interim provision requiring law enforcement officers to conduct background checks was ruled unconstitutional by the Supreme Court. Consequently, gun purchasers were no longer required to fill out the Brady Handgun Purchase Form. However, the five-day waiting period remained in place until November 30, 1998, when it expired. It was replaced by a mandatory computerized criminal background check through the National Instant Criminal Background Check System (NICS) before any firearm purchase from a federally licensed firearms dealer.

People Who May Not Purchase Firearms under the Brady Law

The Brady law prohibits firearms sales to any person who:

- Is charged with a crime punishable by imprisonment for more than one year or has been convicted of such a crime

- Is a fugitive from justice

- Is an unlawful user of a controlled substance

- Has been judged mentally ill or has been committed to a mental institution

- Has renounced U.S. citizenship

- Is subject to a court order restraining him or her from harassing, stalking, or threatening an intimate partner or a child

- Has been convicted of domestic violence

- Is an illegal alien

- Has been dishonorably discharged from the military

"PERMANENT BRADY": THE NICS

On November 30, 1998, the five-day waiting period for handgun purchasers was replaced by the NICS, which was designed to quickly screen purchasers of both handguns and long guns. As a result, all FFLs must verify the identity of a customer and receive authorization for the sale from the NICS, which usually takes less than a minute. This is a permanent provision of the Brady law.

How the NICS Works

Under the NICS, FFLs call a toll-free telephone number and provide information on prospective firearms buyers. The Federal Bureau of Investigation (FBI) checks the NICS, which has access to four databases: an interstate criminal records database containing more than 65 million criminal history records as of December 2013; a database containing information on people who are the subject of protection orders, criminal warrants, and other violations; the NICS's own database, which is not shared with other government agencies and which contained 11.9 million active records as of December 2013; and a database maintained by the U.S. Department of Homeland Security's immigration enforcement section, which allows for background searches of noncitizens who are attempting to receive firearms. From 1999 to 2005 the number of background checks conducted annually under the NICS fluctuated between 8.5 million and 9.1 million. (See Table 3.1.) Starting in 2006, when 10 million checks were conducted, the annual number of checks substantially increased each year, more than doubling by 2013, when nearly 21.1 million background checks were conducted. By June 30, 2014, the NICS had processed 192.4 million background checks since its introduction in 1998.

In many states some or all inquiries are directed to the state's designated point of contact (POC), which will then conduct a background check through the NICS. In states without a designated POC, gun dealers contact the FBI directly, and the FBI runs the check through the NICS. Figure 3.1 shows the components of the national firearm check system and how the NICS fits into the system. (The firearm transferee is the person buying the gun. The federal firearm licensee is the person licensed to sell the gun.) The FFL must initiate a background check on the person wanting to buy a gun. Depending on the state, the FFL does this by contacting either the NICS directly or a state-designated POC. In the POC states, the states act as intermediaries between the FFLs and the NICS.

As of 2014, 36 states, five territories, and the District of Columbia were non-POC entities, which means the FBI conducted all the NICS checks. (See Figure 3.2.) Thirteen states were full POC states, which means they maintained their own Brady NICS Program and conducted their own background checks by electronically accessing the NICS. The rest of the states were partial POC states, in that they conducted NICS checks on handgun transfers and had the FBI conduct checks on long gun transfers.

When the NICS check is initiated, one or more of the following things can happen:

- No disqualifying information is found (the search takes 30 seconds or less) and the transaction can proceed immediately.

- The transaction is briefly delayed because it is necessary for the FBI to look outside the NICS for data.

- The FBI contacts state or local law enforcement for information because it is not possible to conduct an

TABLE 3.1

Total National Instant Criminal Background Check System (NICS) checks, November 30, 1998–June 30, 2014

Year	Jan	Feb	Mar	Apr	May	Jun	Jul	Aug	Sep	Oct	Nov	Dec	Totals
1998											21,196	871,644	892,840
1999	591,355	696,323	753,083	646,712	576,272	569,493	589,476	703,394	808,627	945,701	1,004,333	1,253,354	9,138,123
2000	639,972	707,070	736,543	617,689	538,648	550,561	542,520	682,501	782,087	845,886	898,598	1,000,962	8,543,037
2001	640,528	675,156	729,532	594,723	543,501	540,491	539,498	707,288	864,038	1,029,691	983,186	1,062,559	8,910,191
2002	665,803	694,668	714,665	627,745	569,247	518,351	535,594	693,139	724,123	849,281	887,647	974,059	8,454,322
2003	653,751	708,281	736,864	622,832	567,436	529,334	533,289	683,517	738,371	856,863	842,932	1,008,118	8,481,588
2004	695,000	723,654	738,298	642,589	542,456	546,847	561,773	666,598	740,260	865,741	890,754	1,073,701	8,687,671
2005	685,811	743,070	768,290	658,954	557,058	555,560	561,358	687,012	791,353	852,478	927,419	1,164,582	8,952,945
2006	775,518	820,679	845,219	700,373	626,270	616,097	631,156	833,070	919,487	970,030	1,045,194	1,253,840	10,036,933
2007	894,608	914,954	975,806	840,271	803,051	792,943	757,884	917,358	944,889	1,025,123	1,079,923	1,230,525	11,177,335
2008	942,556	1,021,130	1,040,863	940,961	886,183	819,891	891,224	956,872	973,003	1,183,279	1,529,635	1,523,426	12,709,023
2009	1,213,885	1,259,078	1,345,096	1,225,980	1,023,102	968,145	966,162	1,074,757	1,093,230	1,233,982	1,223,252	1,407,155	14,033,824
2010	1,119,229	1,243,211	1,300,100	1,233,761	1,016,876	1,005,876	1,069,792	1,089,374	1,145,798	1,368,184	1,296,223	1,521,192	14,409,616
2011	1,323,336	1,473,513	1,449,724	1,351,255	1,230,953	1,168,322	1,157,041	1,310,041	1,253,752	1,340,273	1,534,414	1,862,327	16,454,951
2012	1,377,301	1,749,903	1,727,881	1,427,343	1,316,226	1,302,660	1,300,704	1,526,206	1,459,363	1,614,032	2,006,919	2,783,765	19,592,303
2013	2,495,440	2,309,393	2,209,407	1,714,433	1,435,917	1,281,351	1,283,912	1,419,088	1,401,562	1,687,599	1,813,643	2,041,528	21,093,273
2014	1,660,355	2,086,863	2,488,842	1,742,946	1,485,259	1,382,975							10,847,240
Total													192,415,215

Note: These statistics represent the number of firearm background checks initiated through the NICS. They do not represent the number of firearms sold. Based on varying state laws and purchase scenarios, a one-to-one correlation cannot be made between a firearm background check and a firearm sale.

SOURCE: "Total NICS Background Checks: November 30, 1998–June 30, 2014," in *National Instant Criminal Background Check System*, Federal Bureau of Investigation, June 30, 2014, http://www.fbi.gov/about-us/cjis/nics/reports (accessed July 14, 2014)

FIGURE 3.1

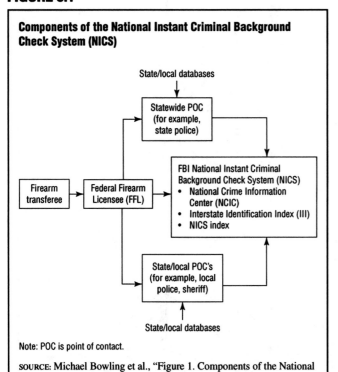

Components of the National Instant Criminal Background Check System (NICS)

Note: POC is point of contact.

SOURCE: Michael Bowling et al., "Figure 1. Components of the National Firearm Check System," in *Background Checks for Firearm Transfers, 2002*, U.S. Department of Justice, Office of Justice Programs, Bureau of Justice Statistics, September 2003, http://bjs.ojp.usdoj.gov/content/pub/pdf/bcft02.pdf (accessed July 14, 2014)

NICS check electronically. The FBI has three business days to complete its background check. If the check cannot be completed in three days, the transaction may still occur, even though potentially disqualifying information might exist in the NICS. This is called a "default proceed" transaction. The dealer does not, however, have to complete the sale, and the FBI will continue to review the case for two more weeks. If disqualifying information is discovered after three business days, the FBI then contacts the dealer to find out whether the gun was transferred (sold) under the default proceed rule. Default proceed transactions are a matter of particular concern because potentially disqualifying information on a gun buyer may well exist. This sometimes occurs when information about a person's recent arrest or conviction has not been entered into the national database.

- The dealer transfers a gun to a prohibited person in a default proceed transaction. If this occurs, the FBI may later notify local law enforcement agencies and the ATF in an attempt to retrieve the gun and take appropriate action, if any, against the buyer.

- The NICS check returns disqualifying information on the buyer and the transfer is denied.

Figure 3.3 illustrates the NICS procedures as they unfolded per 100 applicants in 2012. For every 100 applicants, 70 proceeded immediately and were cleared for purchase, while 30 were transferred to the FBI's NICS section for analysis. Of those 30, 22 were immediately cleared for approval and eight were held for additional review. Ultimately, only 1.01 out of every 100 firearm transactions led to a denial by the NICS.

Table 3.2 shows the reasons for NICS denials of gun transfers between November 30, 1998, and June 30, 2014. The most common reason, which accounted for 56.6% of denials over this period, was the applicant's previous conviction of a felony punishable by more than one year in prison or a misdemeanor punishable by more than two years. Other leading reasons for rejection were an applicant's status as a fugitive from justice (10.6%), an applicant's prior conviction of a misdemeanor crime of domestic violence (9.8%), and an applicant's unlawful use of or addiction to a controlled substance (8.5%).

As of June 30, 2014, there were 11.9 million active records in the NICS system, or 11.9 million individuals known to the FBI to be legally prohibited from owning a firearm. (See Table 3.3.) Over 6 million (50.3%) were ineligible for firearm ownership due to their unlawful immigration status, and 3.5 million (29.5%) were ineligible due to their mental health status. Although people convicted of crimes were the most likely individuals to be denied permits, they represented only 1.7 million (14.2%) active NICS records, suggesting that they were far more likely to attempt to purchase guns illegally than other prohibited people.

Impact of the Brady Law on Gun Violence

Philip J. Cook and Jens Ludwig examine in "The Effects of the Brady Act on Gun Violence" (Bernard E. Harcourt, ed., *Guns, Crime, and Punishment in America*, 2003) the impact of the Brady law on gun violence. They conclude:

> We do not find evidence that the background-check requirement of the Brady Act has reduced lethal violence or affected the choice of weapons by killers. The Brady Act may have reduced firearm suicides among older Americans, the population at highest risk for suicide, although this decline is at least partially offset by some increase in non-gun suicides.... Our analysis of the Brady Act highlights the difficulty of reducing gun violence in America through regulation of the primary market [transactions taking place through FFLs] without some change in policy toward secondary-market transfers [transactions between individuals or between non-FFL dealers and individuals at flea markets and gun shows, which can take place within states that do not regulate such sales].

In "Firearm Death Rates and Association with Level of Firearm Purchase Background Check" (*American Journal of Preventive Medicine*, vol. 35, no. 1, July 2008), Steven A. Sumner, Peter M. Layde, and Clare E. Guse of

FIGURE 3.2

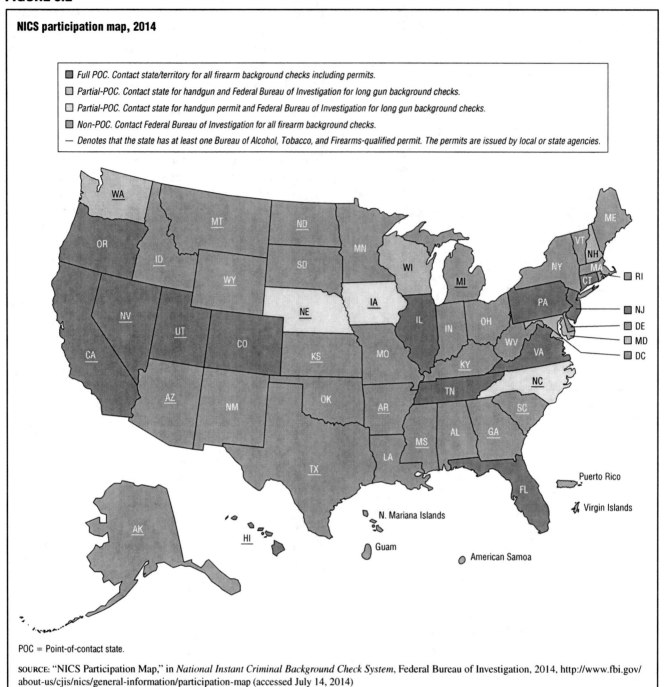

NICS participation map, 2014

■ Full POC. Contact state/territory for all firearm background checks including permits.

□ Partial-POC. Contact state for handgun and Federal Bureau of Investigation for long gun background checks.

□ Partial-POC. Contact state for handgun permit and Federal Bureau of Investigation for long gun background checks.

■ Non-POC. Contact Federal Bureau of Investigation for all firearm background checks.

— Denotes that the state has at least one Bureau of Alcohol, Tobacco, and Firearms-qualified permit. The permits are issued by local or state agencies.

POC = Point-of-contact state.

SOURCE: "NICS Participation Map," in *National Instant Criminal Background Check System*, Federal Bureau of Investigation, 2014, http://www.fbi.gov/about-us/cjis/nics/general-information/participation-map (accessed July 14, 2014)

the Medical College of Wisconsin contend that the agencies conducting background checks under the Brady law often lack the necessary data to conduct a complete search. The researchers suggest that this situation is likely due to multiple factors, including incomplete digitalization of state records, laws that prevent sharing of local data, or budgetary constraints. They determine that states conducting local checks at the time of a firearm purchase showed a lower rate of both homicide and suicide across all age groups. Sumner, Layde, and Guse conclude that "methods to increase local-level agency background checks, such as authorizing local police or sheriff's departments to conduct

them, or developing the capability to share local-level records with federal databases, should be evaluated as a means of reducing firearm deaths."

Controversies over Brady Provisions

Major ongoing Brady law controversies center on four issues:

1. The elimination of the five-day waiting period, beginning on November 30, 1998

2. The length of time the FBI can retain records of gun transactions

FIGURE 3.3

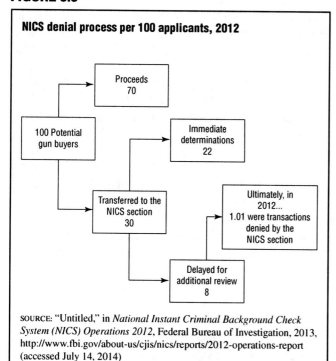

NICS denial process per 100 applicants, 2012

SOURCE: "Untitled," in *National Instant Criminal Background Check System (NICS) Operations 2012*, Federal Bureau of Investigation, 2013, http://www.fbi.gov/about-us/cjis/nics/reports/2012-operations-report (accessed July 14, 2014)

TABLE 3.2

NICS firearm denials, by reason, November 30, 1998–June 30, 2014

Rank	Prohibited category description	Total	Percent of total
1	Convicted of a crime punishable by more than one year or a misdemeanor punishable by more than two years	632,957	56.63%
2	Fugitive from justice	118,837	10.63%
3	Misdemeanor crime of domestic violence conviction	109,247	9.77%
4	Unlawful user/addicted to a controlled substance	95,036	8.50%
5	State prohibitor	53,760	4.81%
6	Protection/restraining order for domestic violence	46,470	4.16%
7	Under indictment/information	26,620	2.38%
8	Adjudicated mental health	14,613	1.31%
9	Illegal/unlawful alien	13,580	1.21%
10	Federally denied persons file	5,696	0.51%
11	Dishonorable discharge	843	0.08%
12	Renounced U.S. citizenship	62	0.01%
	Total federal denials	**1,117,721**	**100.00%**

NICS = National Institute Criminal Background Check System

SOURCE: "Federal Denials," in *National Instant Criminal Background Check System*, Federal Bureau of Investigation, June 30, 2014, http://www.fbi.gov/about-us/cjis/nics/reports (accessed July 14, 2014)

3. The provision of the law that exempts certain people from background checks

4. The regulation of sales at gun shows and on the Internet

ELIMINATION OF THE WAITING PERIOD. The elimination of the waiting period is probably the thorniest issue connected to the Brady law. Gun control advocates

TABLE 3.3

Active records in the NICS Index, June 30, 2014

Rank	Prohibited category description	Total	Percent of total
1	Illegal/unlawful alien	6,006,739	50.30%
2	Adjudicated mental health	3,526,079	29.53%
3	Convicted of a crime punishable by more than one year or a misdemeanor punishable by more than two years	1,697,982	14.22%
4	Fugitive from justice	407,378	3.41%
5	Misdemeanor crime of domestic violence conviction	108,580	0.91%
6	State prohibitor	36,760	0.31%
7	Under indictment/information	33,077	0.28%
8	Federally denied persons file	31,757	0.26%
9	Unlawful user/addicted to a controlled substance	30,270	0.25%
10	Protection/restraining order for domestic violence	27,501	0.23%
11	Renounced U.S. citizenship	25,423	0.21%
12	Dishonorable discharge	10,426	0.09%
	Total active records in the NICS	**11,941,972**	**100.00%**

NICS = National Instant Criminal Background Check System

SOURCE: "Active Records in the NICS Index," in *National Instant Criminal Background Check System*, Federal Bureau of Investigation, June 30, 2014, http://www.fbi.gov/about-us/cjis/nics/reports/nics-index-070314.pdf (accessed July 14, 2014)

support legislation to establish a minimum three-day waiting period. Gun rights advocates claim that 24 hours is adequate. Gun control advocates maintain that a longer waiting period allows local police to search records that might not be in the NICS and that a waiting requirement of at least three days allows a "cooling-off" period, thereby preventing many crimes of passion. They worry that the speed of the background checks may help some of the targets evade the Brady law. This could happen if state and local records have not made it into the national database. For example, the NICS may miss individuals who have only recently been convicted of crimes. However, some states have mandatory waiting periods that must be respected before a person can purchase a firearm.

RETENTION OF RECORDS. Soon after the NICS started operations in 1998, the NRA filed a lawsuit arguing, among other things, that records relating to a background check should be immediately destroyed to prevent the formation of a central registry of gun buyers—a registry the NRA considers an invasion of privacy. The Department of Justice noted that the records would be kept, generally, for 90 days but for no more than six months to conduct audits to make sure gun dealers were following the law and to check for identity fraud and other abuses of the system. In July 2000 the federal court of appeals in the District of Columbia upheld in *NRA v. Reno* (No. 99-5270) the legality of the regulation, allowing the FBI to conduct periodic security audits of NICS records. In June 2001 the Supreme Court declined to hear the NRA's appeal of that decision, thereby allowing the lower court's decision to stand.

Only days later the U.S. attorney general John D. Ashcroft (1942–), a gun rights advocate and the head of the Department of Justice, proposed destroying the critical NICS records within one business day. Congress asked the U.S. General Accounting Office (GAO; now the U.S. Government Accountability Office) to investigate the implications of such a quick destruction of records. In July 2002 the GAO published *Gun Control: Potential Effects of Next-Day Destruction of NICS Background Check Records* (http://www.gao.gov/new.items/d02653.pdf). The GAO stated that 97% of the guns retrieved from illegal purchasers who had been incorrectly approved to buy guns between July 2001 and January 2002 would not have been detected under the attorney general's one-day destruction plan.

Two years later, under an amendment sponsored by Representative Todd Tiahrt (1951–; R-KS), the FBI was mandated to destroy the NICS records of a gun sale within 24 hours of allowing the sale to proceed. The amendment was approved as part of legislation that provided funding for the Departments of Commerce, Justice, and State and made it illegal for the agency to report on sales of multiple handguns and gun trace statistics. The amendment, which included numerous other measures meant to protect the privacy of gun owners and prohibit firearms purchasing records from being used in lawsuits against gun manufacturers, was loosened somewhat in 2008, but the requirement to destroy records within 24 hours remained in effect as of 2014.

EXEMPTION FROM BACKGROUND CHECKS. The Brady law exempts gun buyers from NICS background checks if they have a state permit that meets certain criteria established by the ATF, including holding a right-to-carry permit. However, states that issue permits are expected to incorporate an NICS check into their permit process. Buyers who do not have a permit must have an NICS check, and people renewing their permit must undergo an NICS check at that time.

GUN SHOW AND ONLINE REGULATION. The background check requirement on potential gun buyers applies to federally licensed gun dealers, manufacturers, importers, and pawnshop brokers. People who are not gun dealers and cannot be described as "engaged in the business" of selling firearms do not have to be licensed by the federal government to sell guns, nor do they need to perform background checks on their buyers. This means that someone who is not a gun dealer but who offers a gun or guns for sale at gun shows or flea markets, through classified ads, or online is not subject to the Brady law. According to the Brady Campaign to Prevent Gun Violence, in "Background Checks" (2014, http://www.bradycampaign.org/?q=20-year-anniversary), these unregulated gun sales represent approximately 40% of total gun purchases.

As with sales of many other consumer products, online sales of guns have proliferated in the years since the Brady law was passed. Like gun shows, online marketplaces for guns allow unlicensed individuals to sell guns without conducting background checks; unlike gun shows, online marketplaces never close. Both Craigslist and eBay prohibit sales of firearms and related items, but online gun transactions are common on sites such as Armslist (http://www.armslist.com). Michael Luo, Mike McIntire, and Griff Palmer explain in "Seeking Gun or Selling One, Web Is a Land of Few Rules" (NYTimes.com, April 17, 2013) that "Armslist and similar sites function as unregulated bazaars, where the essential anonymity of the Internet allows unlicensed sellers to advertise scores of weapons and people legally barred from gun ownership to buy them." The reporters analyzed over 170,000 Armslist ads over a period of three months and determined that 94% were posted by private, unlicensed sellers. They also contacted individuals attempting to buy and sell guns on the site. Luo, McIntire, and Palmer found numerous instances in which those intending to buy guns were convicted felons, fugitives from justice, domestic violence offenders, or others prohibited from owning guns. They also found cases in which individuals offered dozens of guns for sale over the three-month period, calling into question their status as individuals not engaged in the business of selling firearms.

In "Study Finds Vast Online Marketplace for Guns without Background Checks" (WashingtonPost.com, August 5, 2013), Philip Rucker notes that a study by the center-left advocacy group Third Way came to similar conclusions. The group analyzed Armslist listings in 10 states whose senators had recently voted against expanding background checks to change the gun show and Internet loophole. Rucker explains, "At any given time, more than 15,000 guns were for sale in those states, according to the study, and more than 5,000 of them were semi-automatic weapons. Nearly 2,000 ads were from prospective buyers asking to purchase specifically from private sellers, where no background checks are required." Rucker quotes Lanae Erickson Hatalsky, the director of social policy at Third Way, who said, "Nobody has any ability to stop these people who are looking for private sellers—and the only reason to do that is to evade the background check system."

The NRA and other gun rights advocates typically contest such descriptions of gun shows and online sales. The Institute for Legislative Action (ILA), the lobbying segment of the NRA, suggests in "2014 NRA-ILA Firearms Fact Card" (March 26, 2014, http://www.nraila.org/news-issues/fact-sheets/2014/2014-nra-ila-firearms-fact-card.aspx?s=&st=&ps=) that universal background checks would not prevent criminals from obtaining firearms because "most criminals obtain guns from theft, the black market, or 'straw purchasers'—people who can pass a background check and who buy guns for criminals." In "Private Sales Restrictions and Gun Registration" (January 17, 2013, http://www.nraila.org/news-issues/fact-sheets/2013/private-sales-restrictions-and-gun-registration.aspx?s=&st=&ps=), the ILA contests the Brady Campaign's finding that 40% of gun purchases are unregulated, maintaining that the actual figure is much lower and that the Brady Campaign's findings are unreliable because the organization surveyed gun owners generally, not armed criminals, who should be the intended targets of any increased gun control efforts.

Mayors against Illegal Guns, an organization of over 1,000 current and former U.S. mayors who advocate for universal background checks, states in "Background Checks" (2014, http://everytown.org/issue/background-checks) that "92 percent of Americans—including 82 percent of gun owners and 74 percent of NRA members—support criminal background checks for all gun sales." Similarly, Lydia Saad of the Gallup Organization reports in *Americans Back Obama's Proposals to Address Gun Violence* (January 23, 2013, http://www.gallup.com/poll/160085/americans-back-obama-proposals-address-gun-violence.aspx) that 91% of Americans support universal background checks.

Legislation seeking to close the gun show and/or Internet loopholes, or to make background checks universal, has been introduced in every session of Congress since 2001, but none of these bills has been passed. In the absence of federal legislation, many states have closed the loopholes through state-level legislation by requiring either universal background checks or expanding background checks to include purchases at gun shows. According to Mayors against Illegal Guns, in "Frequently Asked Questions about Background Checks" (July 2013, http://3gbwir1ummda16xrhf4do9d21bsx.wpengine.netdna-cdn.com/wp-content/uploads/2014/02/MAIG_-_Background_Check_FAQ.pdf), as of 2013 background checks for all handgun sales were required in California, Colorado, Connecticut, Delaware, Iowa, Hawaii, Illinois, Maryland, Massachusetts, Michigan, Nebraska, New Jersey, New York, North Carolina, Pennsylvania, Rhode Island, and the District of Columbia.

THE VIOLENT CRIME CONTROL AND LAW ENFORCEMENT ACT OF 1994
Banning Assault Weapons

Federal laws have banned the possession of automatic-fire guns (machine guns) since 1934 and their importation and manufacture for private use since 1986. A machine gun shoots a stream of bullets when the trigger is pulled, instead of a single shot.

In 1989 a semiautomatic firearm was used by Patrick Purdy (1964–1989) to kill five children at an elementary school in Stockton, California. This type of firearm shoots only one bullet each time the trigger is pulled, but the shooter does not have to do anything to "ready" the next shot; therefore, the shooter can fire as fast as he or she can pull the trigger. Some semiautomatic guns can be converted to automatic firearms.

In response to Purdy's crime, several states passed laws banning the sale and possession of semiautomatic assault-style weapons. It also led to the passage of the Violent Crime Control and Law Enforcement Act of 1994, which included a subsection commonly known as the Semiautomatic Assault Weapons Ban. This act banned the manufacture, transfer, and possession of semiautomatic assault-type firearms but did not outlaw such firearms lawfully possessed before its enactment.

According to the act, banned weapons were defined as guns with a detachable magazine (a cartridge holder that feeds bullets automatically into a gun's chamber) and two or more of the following: a bayonet lug (a metal mount for a thrusting knife), a flash suppressor, a protruding pistol grip, a folding stock (which allows the gun to be stored in a smaller space), or a threaded muzzle (used to attach other devices). The act also banned the making or sale of large-capacity ammunition magazines capable of holding more than 10 rounds (bullets). The federal law listed 19 types of banned semiautomatic firearms, including the Uzi, the TEC-9, the Street Sweeper, and their copycats, all of which are often referred to as "assault weapons." The definition of this term is not clear-cut, but an assault weapon is most frequently defined as a semiautomatic rifle, shotgun, or pistol with a combination of any or all the characteristics and accessories banned by the Violent Crime Control and Law Enforcement Act.

The act exempted at least 650 different sporting rifles. Furthermore, it was legal under this law to buy the accessories to convert these semiautomatic guns to automatic firearms, but the conversion itself was against federal law.

Although the act had several different provisions pertaining to guns, the focus of the debate over its passage was a proposed ban on semiautomatic military-style weapons. Pro-gun supporters repeatedly stressed that the use of the term *assault weapons* to describe the weapons banned under the law was misleading. They complained that the legislation was intended to ban military-style weapons but actually encompassed some ordinary rifles used in hunting and target shooting—weapons that are seldom used to commit crimes. The Semiautomatic Assault Weapons Ban Subsection of the Violent Crime Control and Law Enforcement Act expired on September 13, 2004. Despite broad support for its renewal at that time, it was not renewed; and despite ongoing calls for its reinstatement in the wake of numerous mass shootings carried out with weapons that were formerly banned under the law, it had not been renewed as of 2014. Chapter 9 offers congressional testimony about renewing the Semiautomatic Assault Weapons Ban.

Banning Juveniles from Possessing Handguns or Ammunition

The 1994 Violent Crime Control and Law Enforcement Act also prohibited the possession of a handgun or ammunition by a juvenile under 18 years of age. In addition, it prohibited the sale or private transfer of a handgun or ammunition to juveniles. Exemptions included cases in which the juvenile temporarily used the handgun for employment, with the permission of the owner, such as in ranching or farming, which sometimes require workers to use guns to kill predatory animals that threaten livestock. Other exemptions included using a handgun for target practice, hunting, or a course of instruction on the safe and lawful use of a handgun.

Implementing New Regulations for Obtaining FFLs

Table 3.4 shows the number of active firearms licensees between 1975 and 2013. Even before the changes that came when the 1994 Violent Crime Control and Law Enforcement Act was implemented, the ATF realized that it needed to strengthen the procedures for obtaining an FFL. Between late 1992 and early 1993 news stories, such as Josh Sugarmann's "Gun Market Is Wide Open in America" (CSMonitor.com, April 23, 1993), revealed just how easy it was to get an FFL. The charge for a three-year license was a mere $10, and the only check was a short computer criminal history query.

In the years that followed, the ATF took steps to curtail the number of active FFLs. The Brady law raised the fee to $200 for a three-year license and $90 for a three-year renewal. In March 1994 the ATF sent out new application forms requiring each applicant to submit fingerprint cards and a photograph. These new regulations,

TABLE 3.4

Federal firearms licensees, 1975–2013

Fiscal year	Dealer	Pawn-broker	Collector	Manufacturer of Ammunition	Manufacturer of Firearms	Importer	Destructive device Dealer	Destructive device Manufacturer	Destructive device Importer	Total
1975	146,429	2,813	5,211	6,668	364	403	9	23	7	161,927
1976	150,767	2,882	4,036	7,181	397	403	4	19	8	165,697
1977	157,463	2,943	4,446	7,761	408	419	6	28	10	173,484
1978	152,681	3,113	4,629	7,735	422	417	6	35	14	169,052
1979	153,861	3,388	4,975	8,055	459	426	7	33	12	171,216
1980	155,690	3,608	5,481	8,856	496	430	7	40	11	174,619
1981	168,301	4,308	6,490	10,067	540	519	7	44	20	190,296
1982	184,840	5,002	8,602	12,033	675	676	12	54	24	211,918
1983	200,342	5,388	9,859	13,318	788	795	16	71	36	230,613
1984	195,847	5,140	8,643	11,270	710	704	15	74	40	222,443
1985	219,366	6,207	9,599	11,818	778	881	15	85	45	248,794
1986	235,393	6,998	10,639	12,095	843	1,035	16	95	52	267,166
1987	230,888	7,316	11,094	10,613	852	1,084	16	101	58	262,022
1988	239,637	8,261	12,638	10,169	926	1,123	18	112	69	272,953
1989	231,442	8,626	13,536	8,345	922	989	21	110	72	264,063
1990	235,684	9,029	14,287	7,945	978	946	20	117	73	269,079
1991	241,706	9,625	15,143	7,470	1,059	901	17	120	75	276,116
1992	248,155	10,452	15,820	7,412	1,165	894	15	127	77	284,117
1993	246,984	10,958	16,635	6,947	1,256	924	15	128	78	283,925
1994	213,734	10,872	17,690	6,068	1,302	963	12	122	70	250,833
1995	158,240	10,155	16,354	4,459	1,242	842	14	118	71	191,495
1996	105,398	9,974	14,966	3,144	1,327	786	12	117	70	135,794
1997	79,285	9,956	13,512	2,451	1,414	733	13	118	72	107,554
1998	75,619	10,176	14,875	2,374	1,546	741	12	125	68	105,536
1999	71,290	10,035	17,763	2,247	1,639	755	11	127	75	103,942
2000	67,479	9,737	21,100	2,112	1,773	748	12	125	71	103,157
2001	63,845	9,199	25,145	1,950	1,841	730	14	117	72	102,913
2002	59,829	8,770	30,157	1,763	1,941	735	16	126	74	103,411
2003	57,492	8,521	33,406	1,693	2,046	719	16	130	82	104,105
2004	56,103	8,180	37,206	1,625	2,144	720	16	136	84	106,214
2005	53,833	7,809	40,073	1,502	2,272	696	15	145	87	106,432
2006	51,462	7,386	43,650	1,431	2,411	690	17	170	99	107,316
2007	49,221	6,966	47,690	1,399	2,668	686	23	174	106	108,933
2008	48,261	6,687	52,597	1,420	2,959	688	29	189	113	112,943
2009	47,509	6,675	55,046	1,511	3,543	735	34	215	127	115,395
2010	47,664	6,895	56,680	1,759	4,293	768	40	243	145	118,487
2011	48,676	7,075	59,227	1,895	5,441	811	42	259	161	123,587
2012	50,848	7,426	61,885	2,044	7,423	848	52	261	169	130,956
2013	54,026	7,810	64,449	2,353	9,094	998	57	273	184	139,244

SOURCE: "Exhibit 10. Federal Firearms Licensees Total (1975–2013)," in *Firearms Commerce in the United States: Annual Statistical Update, 2014*, U.S. Department of Justice, Bureau of Alcohol, Tobacco, Firearms and Explosives, April 24, 2014, http://www.atf.gov/sites/default/files/assets/statistics/CommerceReport/firearms_commerce_annual_statistical_report_2014.pdf (accessed July 14, 2014)

as well as an extra step to ensure that the person applying for an FFL complied with state and local laws, were included as part of the Violent Crime Control and Law Enforcement Act. The act also required each licensee to report the theft or loss of a firearm within 48 hours.

There are 11 types of FFLs, including the Type 3 license—collector of curio and relic firearms. Many of the types of FFLs are specific to importers and manufacturers of firearms, their components, and their ammunition. The Type 1 FFL is for a dealer or gunsmith, and the Type 2 FFL is for a pawnbroker who deals in guns.

In *Federal Firearms Licensees: Various Factors Have Contributed to the Decline in the Number of Dealers* (March 1996, http://www.gao.gov/archive/1996/gg96078.pdf), the GAO states that beginning in 1994, due to the revisions in the law, the number of FFLs and the number of FFL applications started to drop sharply. According to the ATF, in *Firearms Commerce in the United States—2001/2002* (2002), this decline is reflected in the number of active FFLs in 1993 (283,193), compared with 1996 (124,286). However, beginning in 1997 the decline appeared to level off, with the overall number of licenses stabilizing between 1997 (106,710) and 2001 (104,840); thereafter, the number of licenses began to increase.

Table 3.5 shows the number of active FFLs by state in 2013. Unsurprisingly, the states with the most FFLs were among the largest U.S. states. Texas (the second-largest state by population) had the most FFLs, at 10,532, followed by California (the largest state), at 8,435, and Florida (the fourth-largest state), at 7,494. New York (the third-largest state) and Illinois (the fifth-largest state) had comparatively few FFLs given their population size, at 4,130 and 5,077, respectively. There were 139,244 FFLs nationally, including those in the District of Columbia and the U.S. territories, in 2013.

Other Provisions of the Violent Crime Control and Law Enforcement Act

The Violent Crime Control and Law Enforcement Act imposed stiffer penalties for using a gun during a violent crime or drug felony. Amendments to the act and policies enacted by the ATF prohibited the possession of firearms by people guilty of domestic abuse. It also tightened rules and regulations for firearms dealers.

GUN LAWS PASSED IN RESPONSE TO SEPTEMBER 11, 2001

Some new gun laws were passed in the wake of the September 11, 2001, terrorist attacks on the United States. In November 2002 Congress passed the Arming Pilots against Terrorism Act, which allows airline pilots to carry handguns in the cockpit. The Law Enforcement Officers Safety Act, signed into law in 2004, allows off-

TABLE 3.5

Federal firearms licensees, by state, 2013

State	FFL population
Alabama	2,404
Alaska	1,011
Arizona	3,062
Arkansas	2,009
California	8,435
Colorado	2,829
Connecticut	1,808
Delaware	345
District of Columbia	23
Florida	7,494
Georgia	3,807
Hawaii	308
Idaho	1,403
Illinois	5,077
Indiana	2,961
Iowa	2,158
Kansas	1,939
Kentucky	2,473
Louisiana	2,157
Maine	3,826
Maryland	3,175
Massachusetts	974
Michigan	4,366
Minnesota	2,722
Mississippi	1,504
Missouri	5,934
Montana	1,524
Nebraska	1,138
Nevada	1,361
New Hampshire	1,176
New Jersey	545
New Mexico	1,150
New York	4,130
North Carolina	4,728
North Dakota	638
Ohio	5,014
Oklahoma	2,461
Oregon	2,608
Pennsylvania	6,227
Rhode Island	602
South Carolina	2,209
South Dakota	798
Tennessee	3,516
Texas	10,532
Utah	1,241
Vermont	558
Virginia	4,441
Washington	2,871
West Virginia	1,469
Wisconsin	3,130
Wyoming	871
Other territories	102
Total	**139,244**

FFL = Federal Firearms Licenses.

SOURCE: "Exhibit 11. Federal Firearms Licenses by State 2013," in *Firearms Commerce in the United States: Annual Statistical Update, 2014*, U.S. Department of Justice, Bureau of Alcohol, Tobacco, Firearms and Explosives, April 24, 2014, http://www.atf.gov/sites/default/files/assets/statistics/CommerceReport/firearms_commerce_annual_statistical_report_2014.pdf (accessed July 14, 2014)

duty or retired police officers with firearms training to travel the country with a concealed weapon.

THE NICS IMPROVEMENT AMENDMENTS ACT OF 2007

The NICS Improvement Amendments Act of 2007 was signed into law by President George W. Bush (1946–) in January 2008. This legislation mandates that information that would prohibit individuals from possessing or purchasing firearms be transmitted from state and local governments and federal agencies to the NICS. States are provided financial assistance to do this work and receive penalties if they fail to comply.

The law was enacted in response to Seung-Hui Cho's (1984–2007) shooting rampage on April 16, 2007, at Virginia Polytechnic Institute and State University (Virginia Tech) in Blacksburg, Virginia. Cho had been ordered by a judge to receive mental health treatment, but that information was never entered into the NICS database. It would have disqualified Cho from purchasing the 9mm semiautomatic handgun and .22-caliber pistol he used to kill 32 students and teachers, and then himself. Although this law had been proposed in Congress since 2002, it languished until the Virginia Tech tragedy brought the issue to wide public attention.

STATE FIREARMS CONTROL LAWS

Laws and constitutional provisions relating to the purchase, ownership, and use of firearms differ dramatically from state to state as well as from county to county and town to town. In most states the right to gun ownership was protected in the state constitution. (See Table 3.6.) As of 2014, only six states—California, Iowa, Maryland, Minnesota, New Jersey, and New York—did not have such a provision. Only 12 states required a permit for the purchase of a firearm: Connecticut, Hawaii, Illinois, Iowa, Maryland, Massachusetts, Michigan, Minnesota, Nebraska, New Jersey, New York, and North Carolina. In five of these states (Connecticut, Hawaii, Illinois, Massachusetts, and New Jersey) permits were required for all guns, including handguns and long guns (shotguns, rifles, and other long guns that fall under the category of "assault weapons"); and in three states (Iowa, Michigan, and Nebraska) permits were required for handguns but not for long guns. In Maryland permits were only necessary for handguns, but long guns falling under the category of "assault weapons" were entirely prohibited; in Minnesota permits were required for handguns and for long guns falling under the category of "assault weapons"; and in New York permits were required only for handguns, but laws specific to New York City required permits for all guns, among other restrictions.

All states had instant background checks under the terms of the Brady law in 2014, but only 11 and the District of Columbia required a waiting period between the purchase and receipt of a firearm: California, Florida, Hawaii, Illinois, Iowa, Maryland, Minnesota, Nebraska, Rhode Island, Washington, and Wisconsin. (See Table 3.6.) Some of these waiting periods varied depending on whether a long gun or handgun was being purchased, and their length varied from 24 hours (for long guns in Illinois) to 14 days (for all guns in Hawaii). Registration (a record of the transfer or ownership of a specific firearm) was required (in various forms) in only six states (California, Connecticut, Hawaii, Maryland, Michigan, and New York) and the District of Columbia.

The Brady Campaign annually publishes state "report cards" that rank states according to the strength or weakness of their gun laws. The state scorecard rankings for 2013 are shown in Figure 3.4. Each state can earn up to 100 points, as in standard academic grading, and the measures evaluated include background checks, reporting of lost or stolen guns, and procedures for keeping guns out of the hands of dangerous people. States with the strongest gun control laws in these and other areas earn the most points, and states with laws considered the most likely to promote firearm violence earn the least points.

The Brady Campaign gave 26 states a failing grade on gun control in 2013. (See Figure 3.4.) Seven states had grades of between D− and D+ and seven states had grades of between C− and C+. Only 10 states had a grade of B− or above: California (A−), Hawaii (B+), Illinois (B), New York (A−), Massachusetts (B+), Rhode Island (B−), Connecticut (A−), New Jersey (A−), Delaware (B−), and Maryland (A−). Among the 10 states with the lowest gun death rate, nine (California, Hawaii, Minnesota, Iowa, New York, New Jersey, Massachusetts, Connecticut, and Rhode Island) received passing grades for the strength of their gun control laws, and one (Maine) received a failing grade. Among the 10 states with the highest gun death rate, nine (Alaska, Montana, Wyoming, Nevada, Arizona, New Mexico, Louisiana, Mississippi, and Tennessee) received failing grades corresponding with the weakest gun control laws, and one (Alabama) was awarded a D−.

Concealed Weapons

According to the ILA, in "Gun Laws" (http://www.nraila.org/gun-laws.aspx), as of 2014 there were 42 right-to-carry (RTC) states, which means those states allowed individuals to carry concealed firearms for protection, either with or without a permit. Of the 42 RTC states, 36 had shall-issue laws. (See Table 3.7.) These laws require local officials to issue a concealed handgun carry permit to anyone who applies, unless the applicant is prohibited by law from carrying a weapon. An additional four states (Alaska, Arizona, Wyoming, and Vermont) did not require a permit for carrying a concealed firearm. Two other RTC states (Alabama and Connecticut) had discretionary-issue rather than shall-issue laws, which means the state government had some discretion over the issuance of permits. In practical terms, however, these states generally grant permits and are therefore considered RTC states.

The remaining eight states were non-RTC states: California, Delaware, Hawaii, Maryland, Massachusetts, New York, New Jersey, and Rhode Island. These states

TABLE 3.6

Firearm purchase laws, by state, 2014

State	Constitutional provision	Permit required to purchase		Waiting period		Registration	
		Long guns	Handguns	Long guns	Handguns	Long guns	Handguns
Alabama	Article 1, Section 26	No	No	None	None	No	No[a]
Alaska	Article 1, Section 19	No	No	None	None	No	No
Arizona	Article 2, Section 26	No	No	None	None	No	No
Arkansas	Article 2, Section 5	No	No	None	None	No	No
California	None	No	No	10 days	10 days	No	Yes
Colorado	Article 2, Section 13	No	No	None	None	No	No
Connecticut	Article 1, Section 15	Yes	Yes	None	None	Yes[b]	No[a]
Delaware	Article 1, Section 20	No	No	None	None	No	No
Florida	Article 1, Section 8	No	No	None	3 days	No	No
Georgia	Article 1, Section 1	No	No	None	None	No	No
Hawaii	Article 1, Section 15	Yes	Yes	14 days	14 days	Yes	Yes
Idaho	Article 1, Section 11	No	No	None	None	No	No
Illinois	Article 1, Section 22	Yes	Yes	24 hours	72 hours	No	No
Indiana	Article 1, Section 32	No	No	None	None	No	No
Iowa	None	No	Yes	None	3 days	No	No
Kansas	Bill of Rights, Section 4	No	No	None	None	No	No
Kentucky	Article 1, Section 1	No	No	None	None	No	No
Louisiana	Article 1, Section 11	No	No	None	None	No[e]	No
Maine	Article 1, Section 16	No	No	None	None	No	No
Maryland	None	No[c]	Yes	None[c]	7 days	No[c]	Yes
Massachusetts	Part 1, Article 17	Yes	Yes	None	None	No[a]	No[a]
Michigan	Article 1, Section 6	No	Yes	None	None	No	Yes
Minnesota	None	Yes[b]	Yes	7 days[b]	7 days	No	No
Mississippi	Article 3, Section 12	No	No	None	None	No	No
Missouri	Article 1, Section 23	No	No	None	None	No	No
Montana	Article 2, Section 12	No	No	None	None	No	No
Nebraska	Article 1, Section 1	No	Yes	None	2 days	No	No
Nevada	Article 1, Section 11	No	No	None	None	No	No
New Hampshire	Part 1, Article 2-a	No	No	None	None	No	No
New Jersey	None	Yes	Yes	None	None	No	No
New Mexico	Article 2, Section 6	No	No	None	None	No	No
New York	None	No[d]	Yes	None	None	No[d]	Yes
North Carolina	Article 1, Section 30	No	Yes	None	None	No	No[a]
North Dakota	Article 1, Section 1	No	No	None	None	No	No
Ohio	Article 1, Section 4	No	No	None	None	No	No
Oklahoma	Article 2, Section 26	No	No	None	None	No	No
Oregon	Article 1, Section 27	No	No	None	None	No	No[a]
Pennsylvania	Article 1, Section 21	No	No	None	None	No	No[a]
Rhode Island	Article 1, Section 22	No	No[e]	7 days	7 days	No	No[e]
South Carolina	Article 1, Section 20	No	No	None	None	No	No
South Dakota	Article 6, Section 24	No	No	None	None	No	No
Tennessee	Article 1, Section 26	No	No	None	None	No	No
Texas	Article 1, Section 23	No	No	None	None	No	No
Utah	Article 1, Section 6	No	No	None	None	No	No
Vermont	Chapter 1, Article 16	No	No	None	None	No	No
Virginia	Article 1, Section 13	No	No	None	None	No	No
Washington	Article 1, Section 24	No	No	None	5 days	No	No[a]
West Virginia	Article 3, Section 22	No	No	None	None	No	No
Wisconsin	Article 1, Section 25	No	No	None	48 hours	No	No
Wyoming	Article 1, Section 24	No	No	None	None	No	No
District of Columbia	N/A	No	No	10 days	10 days	Yes[e]	Yes[e]

Notes:
[a]Though not a complete registry of ownership, varying levels of documentation of transfers or permits are maintained in these states.
[b]For assault weapons.
[c]The purchase or possession of assault weapons is illegal in Maryland.
[d]Stricter gun control laws have been enacted in New York City, including requiring a permit to purchase a rifle or shotgun and restricting permission to carry firearms, even during transport to or from a target range.
[e]Safety course required.

SOURCE: Compiled by Laurie DiMauro and Mark Lane for Gale, © 2014.

generally did not allow handguns to be carried in public areas, but they had very limited and restrictive issue laws that allowed concealed handgun carry permits to be granted in certain circumstances. The District of Columbia prohibited individuals from carrying guns, as did New York City.

In some states the possession of a concealed weapons permit exempts gun purchasers from background check requirements that otherwise apply under the Brady law. Table 3.8 shows state policies regarding concealed weapons permits relative to the Brady background check as of 2014. In Alaska, for example, concealed weapons permits

FIGURE 3.4

The Law Center and Brady Campaign state scorecard, 2013

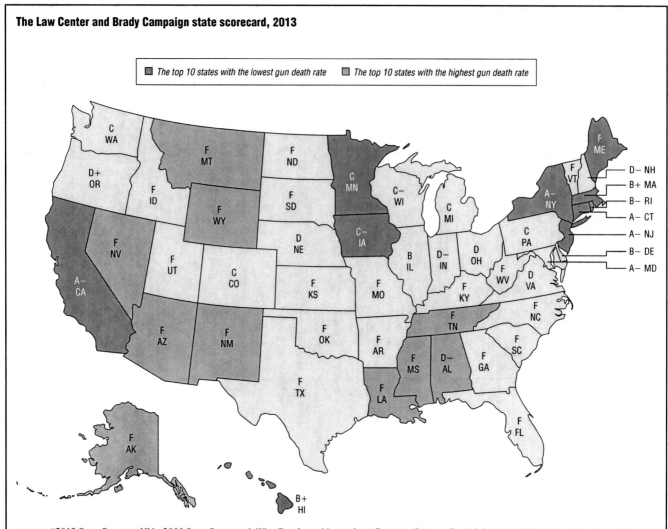

SOURCE: "2013 State Scorecard," in *2013 State Scorecard: Why Gun Laws Matter*, Law Center to Prevent Gun Violence and the Brady Campaign to Prevent Gun Violence, December 9, 2013, http://bradycampaign.org/sites/default/files/SCGLM-Final10-spreads-points.pdf (accessed July 14, 2014)

marked "NICS-exempt" qualify as alternatives to the background check requirements of the Brady law. In other states either a concealed weapons permit or a permit to purchase a handgun after a certain date entitle a purchaser to forgo a new background check.

Supporters of gun rights typically claim that research establishes a positive correlation between the authorization of concealed-carry and decreases in crime. This contention is generally based on the findings of John R. Lott Jr. and David B. Mustard in "Crime, Deterrence, and Right-to-Carry Concealed Handguns" (*Journal of Legal Studies*, vol. 26, no. 1, 1997), and expanded on in Lott's 1998 book *More Guns, Less Crime: Understanding Crime and Gun Control Laws*. Lott analyzed U.S. crime and socioeconomic data and determined that the falling crime rate of the 1990s and beyond was crucially linked to the adoption of shall-issue laws, which have been enacted in 24 states since 1991.

However, this assertion has been contested by other researchers, including a panel assembled by the National Research Council of the National Academies, whose 2004 government-funded report, *Firearms and Violence: A Critical Review*, examined the raw data analyzed by Lott and found that no correlation between gun violence and RTC laws was evident. Numerous other researchers have also attacked Lott's findings. Lott responds to the many criticisms of his work in "What a Balancing Test Will Show for Right-to-Carry Laws" (*Maryland Law Review*, vol. 71, no. 4, 2012), noting that of 29 studies analyzing the issue as of 2012, 18 supported his findings, 10 found that shall-issue laws had no significant effect on crime, and one found that shall-issue laws brought about temporary increases in aggravated assaults.

Open Carry

The phrase "open carry" refers to the carrying of a firearm in plain view while in public. Advocates of open

TABLE 3.7

Right-to-carry weapons laws, 2014

Shall issue	No permit required	Discretionary issue	No right to carry
Arkansas	Alaska	Alabama	California
Colorado	Arizona	Connecticut	Delaware
Florida	Wyoming		Hawaii
Georgia	Vermont		Maryland
Idaho			Massachusetts
Illinois			New York
Indiana			New Jersey
Iowa			Rhode Island
Kansas			
Kentucky			
Louisiana			
Maine			
Michigan			
Minnesota			
Mississippi			
Missouri			
Montana			
Nebraska			
Nevada			
New Hampshire			
New Mexico			
North Carolina			
North Dakota			
Ohio			
Oklahoma			
Oregon			
Pennsylvania			
South Carolina			
South Dakota			
Tennessee			
Texas			
Virginia			
Washington			
West Virginia			
Wisconsin			
Utah			

SOURCE: Created by Mark Lane for Gale, © 2014

TABLE 3.8

Permanent Brady permit chart, 2014

State/territory	Qualifying permits
Alabama	None
Alaska	Concealed weapons permits marked NICS-exempt
American Samoa	None
Arizona	Concealed weapons permits qualify.
Arkansas	Concealed weapons permits issued on or after April 1, 1999 qualify.*
California	Entertainment firearms permit only
Colorado	None
Connecticut	None
Delaware	None*
District of Columbia	None*
Florida	None*
Georgia	Georgia firearms licenses qualify.
Guam	None*
Hawaii	Permits to acquire and licenses to carry qualify.
Idaho	Concealed weapons permits qualify.
Illinois	None
Indiana	None
Iowa	Permits to acquire and permits to carry concealed weapons qualify.
Kansas	Concealed handgun licenses issued on or after July 1, 2010 qualify as alternatives to the background check.
Kentucky	Concealed weapons permits issued on or after July 12, 2006 qualify.
Louisiana	None*
Maine	None*
Maryland	None*
Massachusetts	None*
Michigan	Licenses to purchase a pistol qualify. Concealed pistol licenses (CPLs) issued on or after November 22, 2005, qualify as an alternative to a National Instant Criminal Background Check System (NICS) check. CPLs issued prior to November 22, 2005 and temporary concealed pistol licenses do not qualify as NICS alternative.
Minnesota	None*
Mississippi	License to carry concealed pistol or revolver issued to individuals under Miss. Stat. Ann. § 45-9-101 qualify. (Note: security guard permits issued under Miss. Stat. Ann. § 97-37-7 do not qualify).
Missouri	None*
Montana	Concealed weapons permits qualify.
Nebraska	Concealed handgun permit qualifies as an alternative. Handgun purchase certificates qualify.
Nevada	Concealed carry permit issued on or after July 1, 2011, qualify.
New Hampshire	None
New Jersey	None
New Mexico	None
New York	None
North Carolina	Permits to purchase a handgun and concealed handgun permits qualify.
North Dakota	Concealed weapons permits issued on or after December 1, 1999 qualify.*
Northern Mariana Islands	None
Ohio	None
Oklahoma	None*
Oregon	None*
Pennsylvania	None
Puerto Rico	None

carry, who are among the staunchest defenders of the right to bear arms in the United States, typically promote the practice with the goal of normalizing the presence of guns in public. The practice surged in popularity beginning in 2008, and open-carry demonstrations at political events, restaurants, and other venues have drawn controversy and been extensively covered by the media.

For example, Vanessa O'Connell and Julie Jargon report in "Stores Land in Gun-Control Crossfire" (WSJ.com, March 4, 2010) that as open-carry demonstrations became commonplace, the Starbucks coffee chain initially announced it would not impose any restrictions not mandated by state law. Its locations accordingly became "parade grounds of sorts for open-carry advocates," in the words of O'Connell and Jargon. In 2013 Starbucks reversed its policy, asking that customers not bring firearms into its stores even in open-carry states. The company announced this policy change in "An Open Letter from Howard Schultz, CEO of Starbucks Coffee Company" (September 17, 2013, http://www.starbucks .com/blog/an-open-letter-from-howard-schultz-ceo-of-star bucks-coffee-company/1268), in which Howard Schultz (1953–), the chief executive officer of Starbucks, noted

that the company did not endorse the open-carry movement's portrayal of Starbucks as a "champion of 'open carry'" and expressed dismay at the way the company's

TABLE 3.8

Permanent Brady permit chart, 2014 [CONTINUED]

State/territory	Qualifying permits
Rhode Island	None
South Carolina	Concealed weapons permits qualify.
South Dakota	None*
Tennessee	None
Texas	Concealed weapons permits qualify.
U.S. Virgin Islands	None
Utah	Concealed weapons permits qualify.
Vermont	None
Virginia	None
Washington	Concealed pistol license issued on or after July 22, 2011 qualify.
West Virginia	Concealed handgun license issued on or after June 4, 2014 qualify.
Wisconsin	None
Wyoming	Concealed weapons permits qualify.

*While certain permits issued in these states prior to November 30, 1998, were "grandfathered" as Brady alternatives, none of these grandfathered permits would still be valid under State law as of November 30, 2003.

NICS = National Instant Criminal Background Check System.

CPL = Concealed pistol licenses.

Note: Notwithstanding the dates set forth above, permits qualify as alternatives to the background check requirements of the Brady law for no more than five years from the date of issuance. The permit must be valid under state law in order to qualify as a Brady alternative.

SOURCE: "Permanent Brady Permit Chart," in *Firearms*, U.S. Department of Justice, Bureau of Alcohol, Tobacco, Firearms and Explosives, June 10, 2014, https://www.atf.gov/content/firearms/firearms-industry/permanent-brady-permit-chart (accessed July 15, 2014)

stores were becoming forums for "increasingly uncivil and, in some cases, even threatening" debate on the issue.

Other businesses have faced similar dilemmas. In "Open Carry Group Holds Rally at Home Depot in North Richland Hills" (DallasNews.com, June 1, 2014), Doug J. Swanson reports that the home-improvement chain Home Depot decided in May 2014 to allow open carry within its stores. That same month 150 open-carry activists held a rally outside of a Home Depot store in North Richland Hills, Texas. In contrast, restaurant chains such as Sonic and Chili's, after taking time to review their corporate policies on the issue, eventually settled on a policy of prohibiting open carry in their locations. Other corporations that eventually joined the ranks of businesses asking customers not to openly carry firearms in their stores included the restaurant chain Chipotle and the retail chain Target. Open-carry activists in various states commonly maintain online listings of "OTC-friendly" stores as well as those that prohibit open carry, and they often call for boycotts of such stores.

Federal law does not address open carry, with the exception of laws prohibiting firearms on certain federal properties. According to the Law Center to Prevent Gun Violence, in "Open Carrying Policy Summary" (July 29, 2013, http://smartgunlaws.org/open-carrying-policy-summary), as of 2013 three states (California, Florida, and Illinois) and the District of Columbia prohibit the

open carrying of any weapon, three states (New York, South Carolina, and Texas) prohibit the open carrying of a handgun but not a long gun, and three states (Massachusetts, Minnesota, and New Jersey) prohibit the open carrying of a long gun but not a handgun. Thirteen states require a permit for open carrying of handguns: Connecticut, Georgia, Hawaii, Indiana, Iowa, Maryland, Massachusetts, Minnesota, New Jersey, Oklahoma, Rhode Island, Tennessee, and Utah. Thirty-one states allow open carry of handguns without a permit, although some place certain restrictions on open carry. The open carrying of long guns is allowed without a permit in 44 states, but some states place restrictions on this right. States that allow open carry generally prohibit it in certain locations, including schools, some state-owned properties, bars and other places serving alcohol, and on public transportation.

In most cases the lack of laws regarding open carry is not intentional. Many of these states require a permit to carry concealed weapons, and the fact that the carrying of unconcealed weapons is not regulated is an oversight. Adherents of the open-carry movement typically maintain that the intent of their public behavior is to exercise their right to carry openly so as to ensure that it is not taken away. Additionally, in some states where open carry of handguns is illegal, such as in Texas, open-carry advocates carry long guns in public with the intent of pointing out that the law prevents them from carrying handguns, which would be more practical.

Minors

During the 1990s Americans became concerned with what the U.S. surgeon general called an epidemic of youth violence. In *Youth Violence: A Report of the Surgeon General* (January 2001, http://www.surgeongeneral.gov/library/youthviolence/default.htm), the surgeon general describes the sharply rising arrest rates of young people for violent crimes between 1983 and 1993–94. The surgeon general also notes that juvenile homicides increased 65% between 1987 and 1993. The number of older juveniles killed with firearms accounted for nearly all the growth.

Although arrest rates later declined, the surgeon general explains that some states passed tougher laws prohibiting juveniles from possessing firearms and/or punishing those who supplied them with guns. These state laws are in addition to the federal law that prohibits FFLs from selling or delivering handguns to people under the age of 21 years (18 U.S.C. 922[b][l]) and the federal law that prohibits people under the age of 18 years from possessing handguns (18 U.S.C. Sect. 922[x]). Federal law allows the sale of handguns to people between the ages of 18 and 21 years at gun shows, unless specifically prohibited by state law.

LOCAL ORDINANCES

Many laws regulating guns are local ordinances passed by town, city, and county governments. Most of these statutes regulate the sale of weapons, restricting who can buy firearms and imposing waiting periods on firearms purchases. Among the most notable of these are statutes in effect in New York City, which are more restrictive than the laws in place throughout the rest of the state. For example, the city does not recognize firearm permits from any other jurisdiction, including permits issued by the state of New York. Other restrictions include prohibiting possession of a gun by anyone under 21 years of age, requiring a license to purchase a firearm, and requiring handguns being transported in vehicles in New York City to be unloaded and in a locked container, with ammunition in a separate locked container.

CHAPTER 4
COURT RULINGS ON FIREARMS

The U.S. Constitution and most state constitutions guarantee the right to bear arms, but the courts have ruled that this right may be strictly controlled. Many laws and regulations have been enacted at the local, state, and federal levels to regulate firearms. When these laws have been challenged, state and federal courts have consistently upheld the right of governments to require the registration of firearms, to determine how these weapons may be carried, and even to forbid the sale or use of some weapons under certain circumstances. Courts have also been asked to decide if manufacturers, dealers, or sometimes even relatives of the gun carrier should be held responsible when guns are used to commit crimes.

Table 4.1 presents a list of reasons an individual may be denied the right to bear arms. It has long been generally accepted that gun ownership will be denied to convicted felons, individuals who are not of sound mind, and individuals regarded as incapable due to a mental condition. However, even these restrictions have been challenged in court, and as new prohibited classes of people have been added, there have been further court cases clarifying the prohibitions' constitutionality. The following selection of court cases includes landmark decisions and more recent rulings on gun rights and regulations at the federal, state, and local levels.

SECOND AMENDMENT INTERPRETATIONS
The Bill of Rights: Federal versus State Protections

The "right of the people to keep and bear arms" is the essence of the Second Amendment, which is part of the Bill of Rights (ratified 1791) to the U.S. Constitution. For many years, when deciding cases based on any guarantee granted by the Bill of Rights, including the Second Amendment, the courts relied on the 1833 U.S. Supreme Court decision in *Barron v. City of Baltimore* (32 U.S. 243). In that case the court ruled that the Bill of Rights does not apply to or restrict the states. This meant that the Bill of Rights was a limitation to the federal government only; its protections applied to federal laws.

Under this interpretation, a state did not have to allow people "to keep and bear arms" unless the state's constitution guaranteed that right. If the state constitution did not ensure that right, then state or local authorities could arrest a person who possessed a firearm; nevertheless, federal law enforcement officials could not because of the Bill of Rights. Thus, state constitutions—rather than the Second Amendment of the U.S. Constitution—decided most gun cases. This situation, however, changed during the years of Chief Justice Earl Warren (1891–1974). Warren was the chief justice from 1953 to 1969, when most of the guarantees of the Bill of Rights were held to apply to the states through the 14th Amendment, which states, "No state shall make or enforce any law which shall abridge the privileges or immunities of citizens."

The Individual Rights Interpretation versus the Collective Rights Interpretation

Gun rights advocates interpret the Second Amendment as a guarantee to individuals of the right to keep and bear arms without governmental interference. Favoring stricter control of guns, proponents of the collective rights argument disagree and point to the opening words of the Second Amendment—"A well regulated militia, being necessary to the security of a free state"—as an intention to guarantee the right of states to maintain militias. The modern militia of the United States is the National Guard. Thus, collective rights advocates argue that the general population does not need unfettered access to guns because it is no longer expected to form a militia during times of need.

From 1939 to 2008 legal interpretations of the Second Amendment were based on the Supreme Court's unanimous ruling in *United States v. Miller* (307 U.S. 174 [1939]), which was widely regarded as an unclear opinion. Although many gun control advocates claim the

TABLE 4.1

Firearm prohibitive criteria under NICS

The 11 prohibitive categories for receiving a transferred firearm apply to any individual who:

- Has been convicted in any court of a crime punishable by imprisonment for a term exceeding one year;
- Is a fugitive from justice;
- Is an unlawful user of or addicted to any controlled substance;
- Has been adjudicated as a mental defective or committed to a mental institution;
- Is an alien illegally or unlawfully in the United States or who has been admitted to the United States under a non-immigrant visa;
- Has been discharged from the Armed Forces under dishonorable conditions;
- Having been a citizen of the United States, has renounced U.S. citizenship;
- Is subject to a court order that restrains the person from harassing, stalking, or threatening an intimate partner or child of such intimate partner;
- Has been convicted in any court of a misdemeanor crime of domestic violence; and/or
- Is under indictment for a crime punishable by imprisonment for a term exceeding one year.

NICS = National Institute Criminal Background Check System

SOURCE: "Firearm Prohibitive Criteria," in *NICS Turns 15: Stats Show Success of FBI's Gun Background Check System*, Federal Bureau of Investigation, November 29, 2013, http://www.fbi.gov/news/stories/2013/november/nics-15th-anniversary-stats-show-success-of-gun-background-check-system (accessed July 14, 2014)

ruling declared that the Second Amendment's "obvious purpose" is to "assure the continuation and render possible the effectiveness of" state militias, thus reinforcing a collective rights interpretation, gun rights advocates suggest the ruling established an individual right to bear arms. In any case, during the seven decades that *Miller* served as precedent, lower courts typically upheld the constitutionality of restrictions on the individual's right to bear arms. Thus, although *Miller* has been used to support both sides in the gun control debate, its effect on subsequent legal rulings has been to privilege the collective rights interpretation of the Second Amendment.

The landmark Supreme Court ruling in *District of Columbia v. Heller* (554 U.S. 570 [2008]) established the current status quo in gun control legislation by taking a clear position on the individual right to bear arms. In the *Heller* decision, which concerned a challenge to the District of Columbia's decades-old ban on handguns and its restrictions regarding the storage of firearms in the home, the court unambiguously interpreted the Second Amendment as upholding an individual's right to possess firearms and keep them in the home for the purposes of self-defense. As a result of this new interpretation, the District of Columbia's ban on handguns and its storage requirements for guns were declared unconstitutional. The court did, however, establish the government's right to regulate firearm possession and sales within certain bounds.

FEDERAL COURT CASES

United States v. Cruikshank: Right to Bear Arms?

The first major federal case dealing with the Second Amendment was *United States v. Cruikshank* (92 U.S.

542 [1875]). The defendants were members of the Ku Klux Klan, a white supremacist group, and were convicted of conspiracy to deprive two African American men of their right of assembly and free speech and their right to keep and bear arms as guaranteed by the U.S. Constitution. The Supreme Court ruled, "'Bearing arms for a lawful purpose.' This is not a right granted by the Constitution. Neither is it in any manner dependent upon that instrument for its existence. The second amendment declares that it shall not be infringed; but this ... means no more than that it shall not be infringed by Congress. This is one of the amendments that has no other effect than to restrict the powers of the national government."

In this ruling, the Supreme Court agreed with a lower court ruling that the right to keep and bear arms is a birthright. It is not a right created or conferred by the Constitution. The Constitution, however, guarantees that this right shall not be impaired by the state or federal government. In addition, it is the duty of the state to protect and enforce this right.

United States v. Miller: Possession of Gangster-Type Weapons

The Supreme Court first addressed the meaning of the Second Amendment during the late 1930s, in a case involving a violation of the National Firearms Act of 1934, a federal law designed to make it more difficult to acquire especially dangerous "gangster-type" weapons. Jack Miller and Frank Layton were arrested by federal agents in 1938 and charged with traveling with an unregistered, gangster-type, sawed-off shotgun. A federal district court judge dismissed the case on the grounds that the National Firearms Act violated the Second Amendment. The U.S. government appealed to the Supreme Court in *United States v. Miller*.

The federal government argued that if the Second Amendment protected an individual's right to keep and bear arms, the only arms protected were those suitable to military purposes, not weapons such as sawed-off shotguns that "constitute the arsenal of the 'public enemy' and the 'gangster'"—weapons that the National Firearms Act was intended to regulate.

The Supreme Court reversed the lower court's ruling and upheld the federal law. Because Miller had fled and was not present to plead his case, the court heard only the government's side of the issue and did not hear a strong argument for permitting a citizen to maintain such a weapon. In the end, the Supreme Court denied Miller the right to carry a sawed-off shotgun, noting that no evidence had been presented as to the usefulness "at this time" of a sawed-off shotgun for military purposes. The court stated, "In the absence of any evidence tending to show that possession or use of a 'shotgun having a barrel of less than eighteen inches in length' at this time has some reasonable relationship to the preservation or efficiency of a well regulated militia, we cannot say that the Second Amendment guarantees the right to keep and bear such an instrument. Certainly it is not within judicial notice that this weapon is any part of the ordinary military equipment or that its use could contribute to the common defense."

Referring back to the debates of the Constitutional Convention and the discussion of the militia, the court observed that such deliberations showed "plainly enough that the Militia comprised all males physically capable of acting in concert for the common defense, 'A body of citizens enrolled for military discipline.' And further, that ordinarily when called for service these men were expected to appear bearing arms supplied by themselves and of the kind in common use at the time."

For the next seven decades *Miller* was the major Supreme Court ruling and precedent concerning gun control. Although lower courts typically upheld the constitutionality of placing restrictions on gun ownership using the court's interpretation of the Second Amendment in *Miller*, the ruling has been used to support both sides of the gun rights debate. Gun control advocates have typically interpreted the ruling as establishing only the right of active-duty militia members to keep and bear those arms that are necessary for the common defense, but the court's reasoning is widely considered to be muddled. For example, the court suggested that Miller's right to own a sawed-off shotgun could be constrained by the government, but it also suggested that the government might not have such powers in the case of weapons that could be considered necessary for military use. Some gun rights advocates have maintained that such a case could be made in the latter part of the 20th century, following the use of sawed-off shotguns by U.S. soldiers during the Vietnam War (1954–1975). Because the weapon was shown to have military utility, this line of reasoning holds, individuals have the right to possess them for the sake of the common defense. Furthermore, some have pointed out that the same case can be made for machine guns, which are prohibited under the National Firearms Act, but which have obvious utility on the battlefield.

USING *MILLER* AS A PRECEDENT. The case of *United States v. Tot* (131 F.2d 261 [1942]) originated in the arrest of Frank Tot for stealing cigarettes from an interstate shipment. Tot had previously been convicted of a crime of violence. At the time of his arrest, a .32-caliber Colt automatic pistol was seized during a search of his home. The Third Circuit Court of Appeals did not accept Tot's argument that the Second Amendment prohibited the state of New Jersey from denying him the right to own a gun even if he was a convicted felon. Citing *Miller* as a precedent, the circuit court reasoned that "one could hardly argue seriously that a limitation upon the privilege

of possessing weapons was unconstitutional when applied to a mental patient of the maniac type. The same would be true if the possessor were a child of immature years.... Congress has prohibited the receipt of weapons from interstate transactions by persons who have previously, by due process of law, been shown to be aggressors against society. Such a classification is entirely reasonable and does not infringe upon the preservation of the well regulated militia protected by the Second Amendment."

The circuit court noted that the Second Amendment, "unlike those providing for protection of free speech and freedom of religion, was not adopted with individual rights in mind, but as a protection for the States in the maintenance of their militia organizations against possible encroachments by the federal power."

In 1942 the First Circuit Court of Appeals cited *Miller* in *Cases v. United States* (131 F.2d 916) in an attempt to uphold the Federal Firearms Act of 1938. Jose Cases Velazquez had been convicted of a violent crime and, under the federal law, could not own a gun. The circuit court observed, "The Federal Firearms Act undoubtedly curtails to some extent the right of individuals to keep and bear arms.... [This] is not a right conferred upon the people by the federal constitution."

These rulings (*Tot* and *Cases*), perhaps more clearly than the *Miller* decision itself, established the precedent of interpreting the Second Amendment not as a guarantee of the individual's right to keep and bear arms but as a guarantee of that right so far as was necessary to ensure a well-regulated militia.

Miller was also cited by the Fifth Circuit Court of Appeals in *U.S. v. Emerson* (270 F.3d 203 [2001]). In this case Timothy Joe Emerson was charged with violating the Lautenberg Amendment to the 1994 Gun Act, which prohibits possession of a firearm by people under a domestic violence restraining order. Emerson's estranged wife had obtained such an order from a judge in 1998, after Emerson had acknowledged his mental instability. He was subsequently indicted for illegally possessing two 9mm pistols, a semiautomatic SKS assault rifle with bayonet, a semiautomatic M-14 assault rifle, and an M1 carbine. At his trial in district court, his lawyers argued that the case should be dismissed on the grounds that the federal ban on gun possession by those under a protective order for domestic violence violated the Second Amendment. The district court sided with Emerson and dismissed the charges, reasoning that the provision of the 1994 law violates the Second Amendment because it allows a state court divorce proceeding to deprive a citizen of his or her right to keep and bear arms, even when that citizen has not been found guilty of anything.

In its ruling, the district court noted that it interpreted the Second Amendment as conferring individual rights on U.S. citizens. Nevertheless, U.S. Department of Justice (DOJ) prosecutors appealed the court's decision, stating that it directly conflicted with the long-established legal precedent (the collective rights interpretation) laid down by the Supreme Court in *Miller*.

When the Fifth Circuit Court of Appeals reversed the lower court decision and upheld the domestic violence gun ban against Emerson, gun control advocates viewed the decision as a victory for domestic violence victims and a safeguard for women across the country. Gun rights advocates also found something to praise in the decision, because it seemed to provide support for the argument that individuals are guaranteed the right under the U.S. Constitution to bear arms independent of the provision of a well-regulated militia. The decision of the court stated, "We conclude that *Miller* does not support the government's collective rights or sophisticated collective rights approach to the Second Amendment. Indeed, to the extent that *Miller* sheds light on the matter it cuts against the government's position."

The *Emerson* case sparked conflicting views of the Second Amendment within the DOJ. In arguing the government's case in *Emerson*, the DOJ contended that it is "well settled" that the Second Amendment creates a right held by the states and does not protect an individual's right to bear arms. When Emerson filed his brief on appeal, he attached a copy of a letter from the U.S. attorney general John D. Ashcroft (1942–) to the National Rifle Association of America dated May 17, 2001. The letter stated in part, "Let me state unequivocally my view that the text and the original intent of the Second Amendment clearly protects the right of individuals to keep and bear firearms."

Emerson was quickly seized on by gun rights advocates. In 2000 the attorney Gary Gorski filed a lawsuit in the Eastern District Court in California. The case, *Silveira v. Lockyer* (312 F.3d 1052 [2002]), sought to overturn California's ban on semiautomatic rifles on the basis of the individual right of a person to keep and bear arms under the Second Amendment. The *Silveira* lawsuit lost in the Eastern District Court and was appealed to the Ninth Circuit Court in 2002, which upheld the lower court's decision. The Ninth Circuit Court's written decision strongly rejected the reasoning behind the *Emerson* decision, stating "the debates of the founding era demonstrate that the second of the first ten amendments to the Constitution was included in order to preserve the efficacy of the state militias for the people's defense—not to ensure an individual right to possess weapons."

Silveira has been viewed by some observers as a significant setback for gun rights advocates.

United States v. Synnes: Gun Possession by a Convicted Felon

In *United States v. Synnes* (438 F.2d 764 [1971]), another case involving the possession of a firearm by a convicted felon, the Eighth Circuit Court of Appeals said this about the Second Amendment, "We see no conflict between [a law prohibiting the possession of guns by convicted criminals] and the Second Amendment since there is no showing that prohibiting possession of firearms by felons obstructs the maintenance of a 'well-regulated militia.'"

Most supporters of handgun possession have accepted the right of federal and state governments to deny weapons to former felons, drug abusers, and mentally disabled individuals.

United States v. Warin: Possession of a Machine Gun

In *United States v. Warin* (530 F.2d 103 [1976]), the defendant Francis J. Warin appealed his conviction for possessing an unlicensed submachine gun. Warin tried to convince the Sixth Circuit Court of Appeals that a federal law prohibiting the possession of the gun violated his Second Amendment rights. The court upheld Warin's conviction, stating that it is an "erroneous supposition that the Second Amendment is concerned with the rights of individuals rather than those of the States."

Smith v. United States: Enhanced Penalties for "Use" of Firearms in a Drug Crime

The Supreme Court ruled in *Smith v. United States* (508 U.S. 223 [1993]) that the federal law authorizing stiffer penalties if the defendant "during and in relation to ... [a] drug trafficking crime uses ... a firearm" applies not only to the use of firearms as weapons but also to firearms used as commerce, such as in a bartering or trading transaction. John Angus Smith and a companion went from Tennessee to Florida to buy cocaine, which they planned to resell for profit. During a drug transaction, an undercover agent posing as a pawnshop dealer examined Smith's MAC-10, a compact and lightweight firearm that can be equipped with a silencer and is popular among criminals. Smith told the agent he could have the gun in exchange for 2 ounces (56.7 grams) of cocaine. The officer said he would try to get the drug and return in an hour. In the meantime Smith became suspicious and fled, and after a high-speed chase officers apprehended him.

A grand jury was convened to decide whether there was enough evidence to justify formal charges and a trial. Smith was charged with drug trafficking crimes and with knowingly using the MAC-10 in connection with a drug trafficking crime, among other offenses. Under 18 U.S.C. Section 924(c)(1)(B)(ii), a defendant who uses a firearm in such a way must be sentenced to five years' imprisonment, and if the firearm "is a machine gun or a destructive device, or is equipped with a firearm silencer or firearm muffler," as it was in this case, the sentence is 30 years. Smith was convicted on all counts.

On appeal Smith argued that the law applied only if the firearm was used as a weapon. The 11th Circuit Court of Appeals disagreed, ruling that the federal legislation did not require that the firearm be used as a weapon—"any use of 'the weapon to facilitate in any manner the commission of the offense' suffices." In a similar case, *United States v. Harris* (959 F.2d 246 [1992]), the U.S. Court of Appeals for the District of Columbia Circuit had arrived at the same conclusion. By contrast, the U.S. Court of Appeals for the Ninth Circuit held in *United States v. Phelps* (877 F.2d 28 [1989]) that trading a gun during a drug-related transaction was not "using" it within the meaning of the statute. To resolve the conflict among the different circuit courts, the Supreme Court heard Smith's appeal.

In a 6–3 decision, the court ruled in *Smith v. United States* that the "exchange of a gun for narcotics constitutes 'use' of a firearm 'during and in relation to … [a] drug trafficking crime' within the meaning" of the federal statute. Delivering the opinion of the majority, Justice Sandra Day O'Connor (1930–) wrote that "when a word is not defined by statute, we normally construe it in accord with its ordinary or natural meaning." Definitions for the word *use* from various dictionaries and *Black's Law Dictionary* show the word to mean "convert to one's service; to employ; to carry out a purpose or action by means of." In trying to exchange his MAC-10 for drugs, the defendant "used" or employed the gun as an item of trade to obtain drugs. The phrase "as a weapon" does not appear in the statute. O'Connor reasoned that if Congress had meant the narrow interpretation of "use" (as a weapon only), it would have worded the statute differently.

Justice Antonin Scalia (1936–) dissented from the majority's definition of "use." Defining the normal usage of a gun as discharging, brandishing, or using as a weapon, he observed:

> The Court does not appear to grasp the distinction between how a word can be used and how it ordinarily is used. It would, indeed, be "both reasonable and normal to say that petitioner 'used' his MAC-10 in his drug trafficking offense by trading it for cocaine." … It would also be reasonable and normal to say that he "used" it to scratch his head. When one wishes to describe the action of employing the instrument of a firearm for such unusual purposes, "use" is assuredly a verb one could select. But that says nothing about whether the ordinary meaning of the phrase "uses a firearm" embraces such extraordinary employments. It is unquestionably not reasonable and normal, I think, to say simply "do not use firearms" when one means to prohibit selling or scratching with them.

Bailey v. United States: New Interpretations of "Use"

In 1995 the Supreme Court narrowed the definition of the word *use* that had been established in *Smith*. In *Bailey v. United States* (516 U.S. 137), the court considered the separate criminal misdeeds of Roland J. Bailey and Candisha Robinson.

In 1988 Bailey had been stopped in his car by Washington, D.C., police officers because he was missing a front license plate and an inspection sticker. When Bailey could not produce a driver's license, an officer searched Bailey's car and found ammunition and just over 1 ounce (30 grams) of cocaine. Another officer found a loaded pistol and more than $3,200 in small bills in the trunk. At his trial, Bailey was convicted of possession of cocaine with intent to deliver and using or carrying a firearm in connection with a drug offense. He appealed to the U.S. Court of Appeals for the District of Columbia Circuit.

In June 1991 an undercover police officer approached Candisha Robinson to buy crack cocaine with a marked $20 bill. The officer noticed that she obtained the drugs from her one-bedroom apartment. Later, while executing a search warrant on Robinson's apartment, officers found a .22-caliber derringer, two rocks of crack cocaine, and the marked bill. Robinson was found guilty of cocaine distribution and, among other things, the use or carrying of a firearm during and in relation to a drug trafficking offense. She appealed to the U.S. Court of Appeals for the District of Columbia Circuit.

In Bailey's appeal, *United States v. Bailey* (995 F. 2d 1113 [CADC 1993]), the defense argued there was no evidence that he had used the gun in connection with a drug offense. Robinson argued in her appeal, *United States v. Robinson* (997 F. 2d 884 [CADC 1993]), that during the drug sale to the officer, the gun was unloaded and in a locked trunk and was not used in the commission of or in relation to a drug trafficking offense.

The court of appeals rejected Bailey's claim of insufficient evidence and held that he could be convicted for "using" a firearm if the jury could reasonably infer that the gun had assisted in the commission of a drug offense. In Robinson's case the court reversed her conviction for "using or carrying" because the presence of an unloaded gun in a locked trunk in a bedroom closet was not evidence of actual use. Because the decisions were contradictory, the court of appeals consolidated the two cases and reheard them. A majority of the judges then found that there was sufficient evidence to establish that each defendant had used a firearm in relation to a drug trafficking offense. Bailey and Robinson then jointly appealed to the Supreme Court, which granted their petition to clarify the meaning of "use."

In 1995 the Supreme Court unanimously held that to establish "use," the government must show that the defendant actively employed a firearm so as to make it an "operative factor in relation to the predicate offense." This definition includes hiding a gun in a shirt or pants, threatening to use a gun, or actually using the gun during the commission of a drug crime. The court also found that Bailey's and Robinson's "use" convictions could not be supported because the evidence did not indicate that either defendant actively employed firearms during drug crimes.

In 1998 the Supreme Court put a much broader interpretation on the federal law, which mandates a five-year prison term for a person who "uses or carries" a gun "during and in relation to" a drug-trafficking crime. The three defendants in this case carried guns in the trunks of their cars. The court ruled in *Muscarello v. United States* (524 U.S. 125 [1998]) that having a gun in a car from which a person is dealing drugs fits the meaning of "carries" for purposes of the sentencing statutes.

United States v. Lopez: Possession of a Firearm near a School

The Gun-Free School Zones Act of 1990 made it unlawful for any individual to knowingly possess a firearm in a school zone, defined as within 1,000 feet (305 m) of school grounds, regardless of whether school was in session. Two federal appeals courts came to different conclusions about the constitutionality of the act.

FIFTH CIRCUIT COURT OF APPEALS: THE GUN-FREE SCHOOL ZONES ACT IS UNCONSTITUTIONAL. In March 1992 high school senior Alfonso Lopez Jr. carried a concealed revolver and five bullets into Edison High School in San Antonio, Texas. School officials caught him, and the student was subsequently charged with violating the Gun-Free School Zones Act. Lopez's attorneys moved to dismiss the changes because, they contended, the law was unconstitutional. The trial court did not accept their argument and convicted Lopez. The case then went to the U.S. Court of Appeals for the Fifth Circuit, which disagreed with the trial court, ruling in *United States v. Lopez* (2 F.3d 1342 [1993]) that Congress had exceeded the power granted to it under the commerce clause of the U.S. Constitution when it enacted the Gun-Free School Zones Act. The commerce clause gives Congress the power to regulate conduct that crosses state borders. According to the court, with few specific exceptions, "federal laws proscribing firearm possession require the government to prove a connection to commerce." Congress had made no attempt to link the Gun-Free School Zones Act to commerce in the debates before the law was enacted and in the law itself. The appeals court asserted:

> Both the management of education, and the general control of simple firearms possession by ordinary citizens, have traditionally been a state responsibility.... We are unwilling to ourselves simply assume that the concededly intrastate conduct of mere possession by

any person of any firearm substantially affects interstate commerce, or the regulation thereof, whenever it occurs, or even most of the time that it occurs, within 1,000 feet of the grounds of any school, whether or not then in session. If Congress can thus bar firearms possession because of such a nexus [connection] to the grounds of any public or private school, and can do so without supportive findings or legislative history, on the theory that education affects commerce, then it could also similarly ban lead pencils, "sneakers," Game Boys, or slide rules.

Following this reasoning, the appeals court found the Gun-Free School Zones Act unconstitutional.

NINTH CIRCUIT COURT OF APPEALS: THE GUN-FREE SCHOOL ZONES ACT IS CONSTITUTIONAL. By contrast, the U.S. Court of Appeals for the Ninth Circuit ruled in *United States v. Edwards III* (13 F.3d 291 [1993]) that the commerce clause of the U.S. Constitution did, indeed, give Congress the power to pass a law such as the Gun-Free School Zones Act. In 1991 Sacramento, California, police officers and school officials approached Ray Harold Edwards III and four other males at Grant Union High School. The officers discovered a .22-caliber rifle and a sawed-off rifle in the trunk of Edwards's car. One of the charges against Edwards was violation of the Gun-Free School Zones Act.

Edwards appealed, claiming the law violated the 10th Amendment because Congress did not have the authority under the commerce clause or any other delegated power to enact the Gun-Free School Zones Act. The 10th Amendment states that "the powers not delegated to the United States by the Constitution, nor prohibited by it to the States, are reserved to the States respectively, or to the people."

Disagreeing with *United States v. Lopez*, the court of appeals ruled that the Gun-Free School Zones Act "does not expressly require the Government to establish a nexus between the possession of a firearm in a school zone and interstate commerce.... It is unnecessary for Congress to make express findings that a particular activity or class of activities affects interstate commerce in order to exercise its legislative authority pursuant to the commerce clause."

The court of appeals used *United States v. Evans* (928 F.2d 858 [1991]) as a precedent, which upheld legislation that made it illegal to possess an unregistered machine gun. In *Evans*, the court ruled that "violence created through the possession of firearms adversely affects the national economy, and consequently, it was reasonable for Congress to regulate the possession of firearms pursuant to the commerce clause."

THE U.S. SUPREME COURT: THE GUN-FREE SCHOOL ZONES ACT IS UNCONSTITUTIONAL. In 1995 the Supreme Court struck down the Gun-Free School Zones Act in *United States v. Lopez* (514 U.S. 549) on the grounds that

Congress had overstepped its bounds because it had based the law on the commerce clause of the U.S. Constitution. The commerce clause empowers Congress to regulate interstate commerce, but Congress had failed to connect gun-free school zones with commerce. Chief Justice William H. Rehnquist (1924–2005) wrote that Congress had used the commerce clause as a general police power in a way that is generally retained by states. He also warned that the Gun-Free School Zones Act "is a criminal statute that by its terms has nothing to do with 'commerce' or any sort of economic enterprise, however broadly one might define those terms.... If we were to accept the Government's arguments, we are hard-pressed to posit any activity by an individual that Congress is without power to regulate."

Congress responded in 1996 by approving a slightly revised version of the Gun-Free School Zones Act in the form of amendments to the Department of Defense Appropriations Act of 1997. The amendments required prosecutors to prove an impact on interstate commerce as an element of the offense.

Printz v. United States: The Constitutionality of the Brady Law

Opponents of the Brady Handgun Violence Prevention Act of 1993 challenged its constitutionality soon after it passed. Under the Brady law, Congress had ordered local chief law enforcement officials nationwide to conduct background checks on prospective handgun purchasers who bought their guns through federally licensed dealers. Two sheriffs, Jay Printz of Ravalli County, Montana, and Richard Mack of Graham County, Arizona, charged that Congress exceeded its powers under the 10th Amendment of the U.S. Constitution, which defines the separation of powers—the relationship between the federal government and the sovereign powers of the individual states. They argued that the federal government had placed federal burdens on local police agencies with no federal compensation. Representing the federal government, the U.S. acting solicitor general Walter Dellinger (1941–) argued that the government had the right to require local agencies to carry out federal orders as long as those agencies were not forced to make policy.

In *Printz v. United States* (521 U.S. 898 [1997]), the Supreme Court struck down the Brady law provisions that required local chief law enforcement officials to conduct background checks on prospective handgun buyers and to accept the form on which that background check is based. The court declared that these provisions violated the 10th Amendment to the U.S. Constitution. Justice Scalia wrote, "The Federal Government may neither issue directives requiring the States to address particular problems, nor command the States' officers,

or those of their political subdivisions, to administer or enforce a federal regulatory program."

The court unanimously upheld the Brady law's five-day waiting period for handgun purchases because the waiting period was directed at gun store owners and was not a federal mandate to state officials. Most chief law enforcement officers continued to conduct background checks voluntarily until the National Instant Criminal Background Check System, which was instituted by the Brady law, became effective in November 1998.

Bryan v. United States: Selling Guns without a License

The Firearms Owners' Protection Act of 1986 prohibits any person other than a licensed dealer from dealing in firearms. Anyone who "willfully violates" this law is subject to a fine and can be sentenced up to five years in prison. Sillasse Bryan bought several pistols in Ohio by using straw purchasers (legally qualified buyers who purchase for someone who is not legally qualified). After filing the serial numbers off the guns, he resold the weapons in New York City, in areas known for drug dealing. At his trial the defense argued that Bryan could be convicted only if he knew of the specific federal licensing requirement of the law. His argument failed, and Bryan was convicted. In *Bryan v. United States* (524 U.S. 184 [1998]), the Supreme Court interpreted "willfully violates" to mean that the defendant only needs to know that he was selling guns illegally. Bryan's conviction was upheld.

United States v. Bean: Federal Guns-for-Felons Program

According to the Supreme Court ruling in *United States v. Bean* (537 U.S. 71 [2002]), courts cannot restore firearm privileges to anyone convicted of a felony. In this case Thomas Lamar Bean, a registered gun dealer in Texas, drove to Mexico for dinner with his associates one night after a gun show in 1998. Although he had asked one of his associates to remove the firearms and ammunition from his car, the Mexican police found one box of ammunition remaining inside. Bean was convicted of importing ammunition to Mexico, which is a felony. As a result of his conviction, he was subsequently barred from "possessing, receiving, or distributing firearms or ammunition" in the United States. Under Title 18, Section 925(c), however, Bean was allowed to petition the Bureau of Alcohol, Tobacco, Firearms, and Explosives (ATF) for reinstatement of his gun privileges, known as the guns-for-felons program. The ATF turned him down, stating that it had no money to process his application after the 1992 Appropriations Act passed by Congress had barred the ATF from spending money on such activities. Bean filed suit in the U.S. District Court in Texas, which decided to lift his prohibition. The Fifth Circuit Court of Appeals affirmed that decision.

In 2002 the Supreme Court rejected Bean's contention that the ATF's failure to act amounted to a de facto (existing in reality whether with lawful authority or not) denial of his application. In the decision, Justice Clarence Thomas (1948–) stated that "mere inaction by ATF does not invest a district court with independent jurisdiction to act on an application." The court also ruled unanimously that the federal "relief from disabilities" guns-for-felons program could not be revived by federal judges. Under the "relief from disabilities" program, convicted felons were given the right to apply for "relief" from the "disability" of not being able to buy or possess a gun. The Supreme Court did not comment on Bean's claim that he had a Second Amendment right to get his guns back.

United States v. Stewart: Possession of Homemade Machine Guns

Robert Wilson Stewart Jr. was a convicted felon who sold parts kits to make Maadi-Griffin .50-caliber rifles. He advertised the kits on the Internet and in magazines. The ATF began investigating Stewart when it realized that he had a prior conviction for the possession and transfer of a machine gun. During the investigation an ATF agent purchased parts kits from Stewart and determined that they could be used to make an unlawful firearm. After obtaining a warrant, the ATF searched Stewart's residence and discovered 31 firearms, including five machine guns that Stewart had machined and assembled. Stewart was convicted of being a felon in possession of a firearm, of unlawful possession of a machine gun, and of possessing several unregistered, homemade machine guns.

The case against Stewart had been based, in part, on an interpretation of the commerce clause of the U.S. Constitution that prohibits anyone except a licensed importer, manufacturer, or dealer of firearms to import, manufacture, or deal in firearms. In his appeal, Stewart claimed that Congress had exceeded its commerce clause power and violated the Second Amendment.

In *United States v. Stewart* (348 F.3d 1132 [2003]), the Ninth Circuit Court of Appeals overturned the lower court's ruling on violating the commerce clause, saying that Stewart did not have a substantial effect on interstate commerce. The court, however, affirmed his conviction for being a felon in possession of a firearm.

In 2005 the DOJ appealed the case to the Supreme Court. The High Court would not hear the case but instructed the Ninth Circuit Court of Appeals to further consider the case in light of its recent ruling in *Gonzales v. Raich* (545 U.S. 1 [2005]). That case allowed Congress to use the commerce clause to ban the cultivation and possession of homegrown marijuana for personal medical use because, the court said, the marijuana could affect the supply and demand of the drug, thereby affecting interstate commerce. In June 2006 the Ninth Circuit Court of

Appeals ruled that Congress has the power to regulate the sales of homemade machine guns because they can enter the interstate market and affect supply and demand. Thus, Congress had not exceeded its commerce clause power and had not violated the Second Amendment.

District of Columbia v. Heller: The Constitutionality of the District of Columbia's Handgun Ban

In 2008 the Supreme Court was once again confronted with making a decision that hinged on the Second Amendment and would issue its first decision interpreting the Second Amendment since the 1939 *Miller* decision. This time the interpretation was in the context of the District of Columbia's ban on the possession of handguns. The High Court was expected to make a landmark ruling on the centuries-old question of whether the Constitution's Second Amendment refers to an individual's right to gun ownership or strictly to militia service.

In 1976 the District of Columbia passed a law that banned the private possession of handguns and that required rifles and shotguns in the home to be outfitted with a trigger lock or kept unloaded and disassembled. Dick Anthony Heller was the named party in the suit, but the attorney Robert A. Levy (1941–) personally financed the lawsuit and worked on bringing it to the Supreme Court with the purpose of addressing the Second Amendment question. Attorneys for the District of Columbia contended that the Second Amendment does not give individuals the right to bear arms, whereas Levy and the other attorneys for Heller contended it does.

In 2007 a three-judge panel of the U.S. Court of Appeals for the District of Columbia struck down the gun control ordinance on Second Amendment grounds. Senior Judge Laurence H. Silberman (1935–) wrote for the 2–1 majority that the amendment provides an individual right just as other provisions of the Bill of Rights do. Handguns fall under the definition of "arms"; thus, the District of Columbia may not ban them.

The District of Columbia appealed the case to the Supreme Court. In March 2008 the High Court heard arguments, and in June 2008 the court rendered its decision in *District of Columbia v. Heller*. The 5–4 landmark ruling interpreted the Second Amendment as protecting the individual's right to own a gun, thus supporting the 2007 decision of the U.S. Court of Appeals for the District of Columbia. The majority opinion, written by Justice Scalia, provided for gun control legislation by noting that "like most rights, the Second Amendment right is not unlimited. It is not a right to keep and carry any weapon whatsoever in any manner whatsoever and for whatever purpose." Justice John Paul Stevens (1920–), writing for the minority (dissenting) opinion, noted, "The Court's announcement of a new constitutional right to own and use firearms for private purposes upsets that settled understanding, but leaves for future cases the formidable task of defining the scope of permissible regulations."

McDonald v. Chicago: Must States Follow the Same Second Amendment Rules as the Federal Government?

The District of Columbia is under federal jurisdiction. As a result, the *Heller* ruling does not necessarily apply to state and local jurisdictions. One day after *Heller* was decided, Otis McDonald and three other Chicago residents filed suit in the U.S. District Court for the Northern District of Illinois in an attempt to determine whether states must follow the same Second Amendment rules as federal jurisdictions and to overturn a decades-long ban on handguns in the city of Chicago. In *McDonald v. Chicago* (No. 08 C 3645 [2008]), the district court ruled in favor of the city. The petitioners appealed the case to the U.S. Court of Appeals for the Seventh Circuit. In *McDonald v. Chicago* (567 F 3d 856 [7th Cir. 2009]), the appeals court upheld the decision of the district court, noting that the city's handgun ban did not violate the Constitution because the Supreme Court had not yet declared whether its decision in the *Heller* case established a fundamental right to guns applicable throughout the United States. The Supreme Court agreed to hear the case, *McDonald v. Chicago* (561 U.S. ___ [2010]), which was argued in March 2010. In June 2010 the High Court returned its ruling. In a 5–4 decision, the court stated that the "Second Amendment right is fully applicable to the States."

After *Heller* and *McDonald*

In the years after the landmark *Heller* and *McDonald* rulings many gun rights advocates looked to other cases with the hope that the Supreme Court would clarify further dimensions of their Second Amendment rights. As of November 2014, the most prominent of these cases to have been in contention for the High Court's consideration was *Drake v. Jerejian* (No. 12-1150), a case challenging New Jersey's policy for issuing concealed weapons permits. Unlike shall-issue states, in which government officials are required to issue handgun carry permits to applicants except in extraordinary situations, New Jersey requires citizens to prove a justifiable need for a concealed-carry permit. Many gun rights advocates consider this requirement to be in conflict with the *Heller* ruling establishing the right to self-defense, given the stark contrast between the rights New Jersey gun owners enjoy in their home and the rights they enjoy outside of their home. In May 2014 the court refused to hear the case, which signaled that it was not willing to define a right to self-defense outside of the home.

STATE LAWS

Most state constitutions guarantee the right to bear arms, and this right has either been enacted or strengthened in more than a dozen states since 1970 (see the

TABLE 4.2

U.S. states with laws protecting the right to use deadly force in self-defense, 2014

Alabama	Montana
Alaska	New Hampshire
Arizona	North Carolina
Florida	North Dakota
Georgia	Oklahoma
Idaho	Pennsylvania
Indiana	South Carolina
Kansas	South Dakota
Kentucky	Tennessee
Louisiana	Texas
Maine	Utah
Maryland	West Virginia
Michigan	Wisconsin
Mississippi	Wyoming
Missouri	

SOURCE: Created by Mark Lane for Gale, © 2014

Appendix). Some states clearly tie this right to the militia, whereas other state constitutions and courts have ruled from the perspective self-defense. States that have enacted laws to protect the right to use of weapons, including deadly force, in self-defense are shown in Table 4.2. Two classic examples of cases involving self-defense are *Schubert v. DeBard* (398 NE.2d 1339 [1980]) and *State v. Kessler* (614 P.2d 94 [1980]).

Schubert v. DeBard: The Right to Possess a Handgun for Self-Defense Protected in Indiana

Joseph L. Schubert Jr. wanted a handgun to protect himself from his brother, who he believed was mailing him anonymous threats. Indiana law required that "a person desiring a license to carry a handgun shall apply to the chief of police or corresponding police officer"—in this case Robert L. DeBard, the superintendent of the Indiana State Police. The resulting investigation by DeBard's office found what it considered to be evidence that Schubert was mentally unstable and denied his request. When his application was denied, Schubert filed a petition for review.

Schubert's defense attorney accepted the conclusion of the police investigation that Schubert had some psychological problems, but he argued they were irrelevant to the matter at hand. Article 1, Section 32, of the Indiana constitution guarantees that "the people shall have a right to bear arms, for the defense of themselves and the State." Therefore, the attorney argued, when self-defense was properly indicated as the reason for desiring a firearms license, and the applicant was otherwise qualified, the license could not be withheld because an administrative official had subjectively determined that the applicant's need to defend himself was not justified. (At that point, Schubert had not been found to be mentally incompetent, which is an accepted reason to deny permission to carry a gun.)

The Third District Court of Appeals of the State of Indiana agreed with Schubert. After studying the debates surrounding the creation of the Indiana constitution in 1850, most of the appeals court judges concluded, "We think it clear that our constitution provides our citizenry the right to bear arms for their self-defense." If it were left to a police official to determine a "proper reason" for a person to claim self-defense, "it would supplant a right with a mere administrative privilege." Based on this conclusion, the court sent the case back to the lower court, asking it to determine if Schubert was mentally incompetent, which was an accepted basis for denying him the right to purchase a weapon.

Doe v. Portland Housing Authority: Maine State Law Preempts the Portland Public Housing Authority Provision against Gun Possession

Most state constitutions guarantee the right to bear arms, but state laws regulate their possession. The case of *Doe v. Portland Housing Authority* (656 A.2d 1200 [1995]) is an example of such regulation and its preemption (taking precedence) of a public housing authority provision.

A Maine couple identified as John and Jane Doe, who had lived in public housing since 1981, were threatened with eviction after the Portland Housing Authority (PHA) in Maine discovered guns in their apartment. John Doe was a veteran of the U.S. Marine Corps, a former firearms dealer, and a licensed hunter. Jane Doe, a target shooter, reported that she kept a handgun for self-protection when her husband worked late. The Does filed a petition to prevent the PHA from enforcing a provision in their lease that banned the possession of firearms.

The Does argued that a state law that regulates the possession of firearms preempted the lease. The Maine preemption statute declares, "The State intends to occupy and preempt the entire field of legislation concerning the regulation of firearms.... No political subdivision of the State, including, *but not limited to*, municipalities, counties, townships and village corporations, may adopt any ... [law] concerning ... firearms, components, ammunition or supplies."

The PHA claimed that the state law could not preempt its resolutions because the PHA is not a political subdivision listed in the statute. In 1995 the Maine Supreme Court ruled in *Doe v. Portland Housing Authority* that the PHA was indeed a political subdivision. The court also found that the state legislature intended to regulate uniformly the possession of firearms by all Maine residents whether they live in public housing or not. The case was appealed to the U.S. Supreme Court, but the court declined to hear it.

State v. Owenby: Mental Illness Limits the Right to Bear Firearms in Oregon

In 1991 the Circuit Court of Multnomah County, Oregon, ruled that Patrick Owenby, who suffered from mental illness and had carefully planned a murder, was a danger to himself and others. Because Owenby was unwilling, unable, or unlikely to seek voluntary treatment, the court had him committed to a psychiatric facility. It then ordered that Owenby be prohibited from purchasing or possessing firearms for a period of five years, in accordance with Oregon statutes. Owenby appealed the case.

The Oregon Court of Appeals ruled in *State v. Owenby* (111 Or. App. 270, 826 P.2d 51 [1992]) that the statute in question was a narrowly drawn and reasonable restriction on the right to bear arms, was not in violation of the state constitution, was supported by clear and convincing evidence, and did not violate the federal due process clause as expressed in the 14th Amendment: "Nor shall any State deprive any person of life, liberty, or property, without due process of law."

The court stated:

> The right to bear arms is not absolute. In the exercise of its police power, the legislature may enact reasonable regulations limiting the right and has done so.

> Given the nature of firearms ... the danger that the statute seeks to avert is a serious one. The restriction on the right of mentally ill people to bear arms, on the other hand, is a relatively minor one. The statute is narrowly drawn and may be invoked only when it is shown that the prohibition is *necessary* "as a result of the mentally ill person's mental or psychological state," as demonstrated by past behavior that involves unlawful violence.

Benjamin v. Bailey: Ban on Semiautomatic Firearms in Connecticut Upheld

In 1995 the Connecticut Supreme Court upheld in *Benjamin v. Bailey* (234 Conn. 455, 662 A.2d 1226) a 1993 state law banning the sale, possession, or transfer of 67 types of automatic and semiautomatic or burst-fire firearms, ruling that the ban did not violate the state constitutional right to bear arms. The decision made Connecticut one of the first states to have an assault-weapons ban pass legal challenge even though the right of self-defense was specified in its constitution.

State v. Wilchinski: Child Access Prevention in Connecticut Challenged

Florida was the first state to pass a child access prevention law (1989). Often referred to as the safe-storage law, it requires adults to either keep loaded guns in a place reasonably inaccessible to children or use a device to lock the gun. If a child (defined as anyone under the age of 16 years) obtains an improperly stored and loaded gun, the adult owner is held criminally liable.

According to the Law Center to Prevent Gun Violence, in "Child Access Prevention Policy Summary" (August 1, 2013, http://smartgunlaws.org/child-access-prevention-policy-summary), as of 2013, 28 states and the District of Columbia had passed child access prevention laws.

Among these states is Connecticut, whose law was challenged by Joseph Wilchinski, a police officer employed by Central Connecticut State University. He was charged with criminal negligence and sentenced to three years' probation after his teenage son and another boy found a loaded revolver in Wilchinski's bedroom in 1993. The boy was shot and died two days later. Wilchinski appealed his conviction, claiming the law was unconstitutionally vague.

In 1997 the Connecticut Supreme Court upheld the law, declaring in *State v. Wilchinski* (242 Conn. 211) that the requirement to store firearms in a "securely locked box or other container or in a location which a reasonable person would believe to be secure" was sufficiently clear to inform Wilchinski of safe-storage practices.

American Shooting Sports Council, Inc. v. Attorney General: A Challenge to Gun Safety Regulations in Massachusetts

Regulations applying consumer product safety guidelines to all handguns made or sold within the state of Massachusetts were established in October 1997 by Scott Harshbarger (1941–), the state attorney general. They rank among the nation's strongest gun-safety regulations. The day before the rules took effect, the American Shooting Sports Council and a group of Massachusetts gun manufacturers sued to block them, arguing that the attorney general had exceeded his authority. The case ultimately made its way to the Supreme Judicial Court of Massachusetts, which reversed the trial court's ruling in favor of the gun manufacturers. The matter was then sent back to the trial court, where a final ruling in favor of the attorney general was entered in *American Shooting Sports Council, Inc. v. Attorney General* (429 Mass. 871, 711 N.E.2d 899 [1999]). On April 3, 2000, Harshbarger announced that the regulations were in effect immediately.

Coalition of New Jersey Sportsmen v. Whitman: New Jersey's Assault Weapons Ban Challenged

A group of gun clubs and arms manufacturers sought to overturn New Jersey's 1999 ban on assault weapons on the grounds of vagueness, free speech, and equal protection. The U.S. District Court for the District of New Jersey rejected their challenge in March 1999. The court held in *Coalition of New Jersey Sportsmen v. Whitman* (44 F. Supp. 2d 666) that the statute banning assault weapons was not vague because it "addresses an understandable core of banned guns and adequately puts gun owners on notice that their weapon could be prohibited."

In the ruling, the court further held that the statute's ban on specifically named weapons "does not violate anyone's free speech," nor does the statute infringe on equal protection rights "because the rationality of the link between public safety and proscribing assault weapons is obvious." The federal court's decision was affirmed in March 2001 by the U.S. Court of Appeals for the Third Circuit (No. 99-5296).

Oklahoma: Guns on Corporate Property Challenged

On November 1, 2004, amendments to the Oklahoma Firearms Act and the Oklahoma Self-Defense Act took effect, allowing guns in locked vehicles on corporate property in Oklahoma. The amendments were passed after 12 workers at an Oklahoma Weyerhaeuser paper mill were fired for violating a company ban on firearms in the company parking lot. Whirlpool Corporation, Williams Companies Inc., and ConocoPhillips filed a lawsuit against the state in federal court, arguing that the amendments were unconstitutional and prevented them from banning firearms in their parking lots to help ensure safe workplaces. A U.S. district court judge issued a temporary restraining order to prevent the amendments from going into effect until the courts made a final ruling.

In March 2005 the Oklahoma Court of Criminal Appeals ruled that the amendments were criminal in nature, rather than civil. This ruling was necessary to guide the U.S. district court judge in his determination. In October 2007 Terence C. Kern (1944–), the U.S. district judge, placed a permanent injunction against the Oklahoma amendments, ruling that they were in conflict with federal safety laws meant to protect employees at their jobs, primarily denoted in the Occupational Safety and Health Act of 1970 (OSH Act), which created the Occupational Safety and Health Administration (OSHA). However, the issue remained in dispute, and in February 2009 the U.S. Court of Appeals for the 10th Circuit lifted the injunction in *Ramsey Winch Inc. v. Henry* (555 F.3d 1199). In its ruling, the court found:

> The district court held that gun-related workplace violence was a "recognized hazard" under the general duty clause, and, therefore, an employer that allows firearms in the company parking lot may violate the OSH Act. We disagree. OSHA has not indicated in any way that employers should prohibit firearms from company parking lots. OSHA's website, guidelines, and citation history do not speak at all to any such prohibition. In fact, OSHA declined a request to promulgate a standard banning firearms from the workplace.... In declining this request, OSHA stressed reliance on its voluntary guidelines and deference "to other federal, state, and local law-enforcement agencies to regulate workplace homicides."

Numerous states followed Oklahoma's lead after 2009. According to Sara Murray, in "Guns in the Parking Lot: A Delicate Workplace Issue" (WSJ.com, October

15, 2013), as of late 2013, 22 states had passed so-called Bring Your Guns to Work laws, whereby employees were entitled to keep guns in their locked private vehicles on company property. "Some companies have taken the changes in stride," Murray writes, "but others are rewriting their human-resources policies, training employees to detect early signs of employee aggression and considering extra security for tense situations like termination meetings. Law firms specializing in labor and employment say managers are bombarding them with questions about adapting to the new measures."

Paula Fiscal et al. v. City and County of San Francisco et al.: California State Law Preempts San Francisco Ordinance That Bans Handguns in the City

The 2008 California Supreme Court ruling in *Paula Fiscal et al. v. City and County of San Francisco et al.* (70 Cal.Rptr.3d 324) is an example of state law preempting a local ordinance. California state law regulates firearms within California, including their manufacture, distribution, sale, possession, and transfer. In November 2005 San Francisco voters passed Proposition H, a citywide ordinance that would ban the manufacture, distribution, sale, and transfer of firearms and ammunition within San Francisco, as well as prohibit San Francisco residents from possessing handguns within the city.

A few days after Proposition H was passed by city voters, the National Rifle Association of America and the Second Amendment Foundation, a nonprofit group that promotes the right to bear arms, filed suit to block the ordinance. In June 2006 Judge James Warren of the San Francisco Superior Court struck down the ordinance, stating that its key aspects were preempted by state law. He held that under California law local officials cannot ban the possession of firearms from law-abiding citizens.

The city appealed the superior court decision, and in January 2008 a three-judge panel of the California Court of Appeals unanimously ruled to uphold the lower court's decision. The city then appealed the decision to the California Supreme Court, which upheld the previous two decisions. The April 2008 decision by California's high court exhausted the city's possibilities to appeal the case further.

LOCAL RULINGS

Portland, Oregon: Possession of Guns Regulated

The city of Portland, Oregon, passed an ordinance prohibiting "any person on a public street or in a public place to carry a firearm upon his person, or in a vehicle under his control or in which he is an occupant, unless all ammunition has been removed from the chamber and from the cylinder, clip, or magazine" (Portland Ordinance, PCC 14.138210). In 1982 Michael Boyce was convicted of violating this statute. He appealed, contending that the law violated Article 1, Section 27, of the

Oregon constitution, which states, "The people shall have the right to bear arms for the defence of themselves, and the State, but the Military shall be kept in strict subordination to the civil power."

Boyce based his case on *State v. Kessler* (289 Or. 359 [1980]), in which the Oregon Supreme Court upheld the right to possess a billy club or any other type of small weapon for self-defense, and *State v. Blocker* (291 Or. 255, 630 P.2d 824 [1981]), in which the Oregon Supreme Court declared unconstitutional an Oregon law banning a number of weapons, including switchblades, billy clubs, and blackjacks. The court of appeals did not, however, see the similarities and upheld in *State v. Boyce* (658 P.2d 577 [1983]) the lower court's conviction. The court of appeals observed that the statute in *Kessler* and *Blocker* forbids the "mere possession" of certain weapons and that was the characteristic that made it unconstitutional. The statute in this case regulates only the manner of possession, something both *Kessler* and *Blocker* recognized as permissible when the regulation was reasonable. As such, the city of Portland could regulate the use of weapons within its borders.

In fulfilling its obligation to protect the health, safety, and welfare of its citizens, a government body must sometimes pass legislation that touches on a right guaranteed by the state or federal constitution. Such an encroachment is permissible when the unrestricted exercise of the right poses a clear threat to the interests and welfare of the public in general, and the means chosen by the government body do not unreasonably interfere with the right.

The court of appeals agreed that individuals had a right to protect their property and themselves:

> when a threat to person or property arises in the victim's defense capacity. It is true, on the other hand, that, when the threat arises in a public place, the fact that a person must have any ammunition separated from his firearm will hinder him to the extent that he is put to the trouble of loading the weapon.
>
> However, given the magnitude of the city's felt need to protect the public from an epidemic of random shootings, we think that the hindrance is permissible.

Renton, Washington: Guns Not Permitted Where Alcohol Is Served

The city of Renton, Washington, enacted Municipal Ordinance 3477-59, which states, "It is unlawful for anyone on or in any premise in the City of Renton where alcoholic beverages are dispensed by the drink to ... carry any rifle, shotgun, or pistol, whether said person has a license or permit to carry said firearm or not, and whether said firearm is concealed or not."

Four residents, with the support of the Second Amendment Foundation, went to court seeking an injunction on the ordinance, claiming it violated state law and was unconstitutional. The Superior Court of King County upheld the city ordinance, so the Second Amendment Foundation took the case to the Court of Appeals of Washington.

Article 1, Section 24, of the Washington constitution states, "The right of the individual citizen to bear arms in defense of himself, or the state, shall not be impaired, but nothing in this section shall be construed as authorizing individuals or corporations to organize, maintain or employ an armed body of men."

The court indicated in *Second Amendment Foundation v. City of Renton* (668 F.2d 596 [1983]) that "it has long been recognized that the constitutional right to keep and bear arms is subject to reasonable regulation by the State under its police power." It also explained that simply because a right is guaranteed by either the state or federal constitution does not mean that it cannot be regulated. According to the court:

> The scope of permissible regulation must depend upon a balancing of the public benefit to be derived from the regulation against the degree to which it frustrates the purposes of the constitutional provision. The right to own and bear arms is only minimally reduced by limiting their possession in bars. The benefit to public safety by reducing the possibility of armed conflict while under the influence of alcohol outweighs the general right to bear arms in defense of self and state.

> … On balance, the public's right to a limited and reasonable exercise of police power must prevail against the individual's right to bear arms in public places where liquor is served.

Furthermore, the court noted that the statutes "do not expressly state an unqualified right to be in possession of a firearm at any time or place." Had the city of Renton instituted "an absolute and unqualified local prohibition against possession of a pistol by the holder of a state permit," it would have conflicted with state law and Washington's constitution. This it did not do. Rather, the city had instituted a law "which is a limited prohibition reasonably related to particular places and necessary to protect the public safety, health, morals, and general welfare."

Finally, the Court of Appeals of Washington noted that "while 36 states have constitutional provisions concerning the right to bear arms, in none is the right deemed absolute." Furthermore, "those states with constitutional provisions similar to ours (Alabama, Michigan, Wyoming, Oregon, [and] Indiana) have uniformly held the right subject to reasonable exercise of the police power." The city of Renton was within its rights when it passed the ordinance barring firearms from bars, and the court upheld the decision of the lower court.

Morton Grove, Illinois: Handguns Are Banned

As soon as Morton Grove, Illinois, a Chicago suburb, passed an ordinance banning handguns in 1981, handgun owners challenged the city in court. Article 1, Section 22, of the Illinois constitution provides that "subject only to the police power, the right of the individual citizen to keep and bear arms shall not be infringed." Handgun owners argued that the right to bear arms was protected by state and federal constitutions and further contended that if Morton Grove were allowed to pass such laws in contradiction to other towns and cities, a "patchwork quilt" situation would result, in that handgun owners would never know if they were violating a law when traveling from town to town.

Morton Grove defended itself, claiming it was within its power to limit or ban the possession of handguns if city officials believed handgun possession was a threat to peace and stability. The city further claimed that its ordinance did not violate Section 22 of the Illinois constitution because it guaranteed the right to keep "some guns." The Morton Grove law did not ban all guns, only handguns.

LEGAL PROCEEDINGS. The cases of Victor Quilici, Robert Stengl, George Reichert, and Robert Metler were combined and brought to the Federal District Court of Northern Illinois in *Quilici v. Village of Morton Grove* (532 F.Supp. 1169 [1981]), in which the court upheld the town's right to ban handguns. The U.S. Court of Appeals for the Seventh Circuit also upheld in *Quilici v. Village of Morton Grove* (695 F.2d 261 [7th Cir. 1982]) the findings of the district court, saying "the right to keep and bear handguns is not guaranteed by the Second Amendment." The case was appealed to the U.S. Supreme Court. The High Court refused to hear the case, so the ruling of the lower court of appeals stood.

A NEW ROUTE. The Morton Grove handgun owners then went to the Circuit Court of Cook County for an injunction to prevent Morton Grove from instituting the ordinance that banned handguns. The county circuit court upheld the validity of the ordinance. The handgun owners next appealed to the Appellate Court of Illinois, First District, Third Division. In *Kalodimos v. Village of Morton Grove* (447 NE.2d 849 [1983]), the court upheld the decisions of the lower courts. Although the court agreed with the handgun owners that "gun control legislation could vary from municipality to municipality, we find that the framers [of the Illinois constitution] envisioned this kind of local control."

The case was again appealed, and in October 1984 the Illinois Supreme Court upheld in *Kalodimos v. Village of Morton Grove* (53 LW 2233) the lower courts' decisions. Agreeing with earlier observations, the state's highest court noted that "while the right to possess firearms for the purposes of self-defense may be necessary to protect important personal liberties from encroachment by others, it does not lie at the heart of the relationship between individuals and their government." Thus, Morton Grove needed only to have had a "rational basis" for instituting its ban on handguns. The Illinois Supreme Court concluded, "Because of the comparative ease with which handguns can be concealed and handled, a ban on handguns could rationally have been viewed as a way of reducing the frequency of premeditated violent attacks as well as unplanned criminal shootings in the heat of passion or an overreaction to fears of assault, accidental shootings by children or by adults who are unaware that a handgun is loaded, or suicides. The ordinance is a proper exercise of the police power."

Chicago: Limits on Handgun Possession

On March 19, 1982, the Chicago City Council passed an ordinance prohibiting the registration of any handgun after April 10, 1982, the effective date of the ordinance, unless it was "validly registered to a current owner in the City of Chicago" before April 10, 1982 (*Municipal Code of the City of Chicago*, Chapter 8-20-050[c][1]). Jerome Sklar lived in neighboring Skokie, Illinois, when the law was passed. He owned a handgun and held a valid Illinois Firearms Identification Card. On April 15, 1982, after the ordinance had gone into effect, he moved to Chicago. He could not register the weapon and, therefore, was unable to bring it into the city.

Sklar went to court, claiming that the city of Chicago had violated the equal protection clause of the U.S. Constitution because he was unable to register the gun that he owned, whereas owners of firearms who resided in Chicago before the effective date of the ordinance had an opportunity to take advantage of the law's registration requirements. By this time, *Quilici v. Village of Morton Grove* had been decided by the Seventh Circuit Court of Appeals, a decision that applied to this judicial region. Therefore, the U.S. District Court for the Northern District of Illinois indicated in *Sklar v. Byrne* (727 F.2nd 633 [1983]) that because of the *Quilici* ruling it "concluded that the Chicago firearms ordinance does not infringe any federal constitutional right." The court indicated that the city of Chicago had legitimately and rationally used its police power to promote the health and safety of its citizens. Sklar's argument that Chicago could have chosen better ways to protect its citizens from the negative effects of firearms was irrelevant.

The court concluded that the ordinance did not violate the equal protection clause of the U.S. Constitution by limiting new registrations instead of banning handguns altogether. The city was under no legal requirement to take an all or nothing approach to limiting handguns.

Sklar appealed the district court's decision to the U.S. Court of Appeals for the Seventh Circuit, the same court that had ruled on *Quilici v. Village of Morton Grove*. In his appeal, Sklar claimed that his constitutionally guaranteed right to travel had been violated because he could not move into Chicago without giving up his gun. In *Sklar v. Byrne* (727 F.2d 633 [1984]), the court upheld the lower court's decision, citing the precedent established in *Quilici*. The court did not believe that a fundamental constitutional issue was involved. Therefore, the city of Chicago had a right to institute local regulations as long as it did not go overboard. The court stated:

> The Chicago handgun ordinance as a whole promotes legitimate government goals. The city council set forth its purposes in the preamble to the ordinance. The council found that handguns and other firearms play a major role in crimes and accidental deaths and injuries, and that the "convenient availability" of firearms and ammunition contributed to deaths and injuries in Chicago. The council therefore enacted the ordinance to restrict the availability of firearms and thereby to prevent some deaths and injuries among Chicago citizens. The city's primary goals are thus classic examples of the city's police power to protect the health and safety of its citizens.

Sklar argued that it was irrational and "inconsistent with the overall purposes of the ordinance" to allow some people to have handguns and others not to, and not to classify gun owners on the basis of their ability to handle handguns safely. In dismissing his claim, the court stated, "that argument essentially asks this court to second-guess the judgment of the city council. The Constitution does not require the city council to act with a single purpose or to be entirely consistent. Indeed, the council is a political body for the accommodation of many conflicting interests.... The Constitution does not require the city council to enact the perfect law. The council may proceed step by step, 'adopting regulations that only partially ameliorate a perceived evil and deferring complete elimination of the evil to future regulations.'"

In both *Quilici* and *Sklar*, the courts were not saying that handgun control is or is not a good decision for any local authority to make. They did not see the possession of a handgun as a fundamental right protected by either the federal or the state constitution. The courts stated that a town or city, under the police powers granted it by American tradition and the Illinois constitution, has the right to decide and implement such an ordinance for its own people. A local ordinance does not have to be consistent, as long as the city council can prove that it thought out its decision rationally.

As noted earlier, the U.S. Supreme Court held in *McDonald v. Chicago* that the Second Amendment right of individuals to keep and bear arms for self-defense may not be preempted by state or local laws. The High Court found that the restrictions on gun ownership in Chicago and neighboring Oak Park, Illinois, were too restrictive because they denied citizens the right to legally possess a gun in their own home for self-defense. The court left open the possibility, however, that some state and local gun control measures could withstand constitutional scrutiny applied in the case. Writing for the majority, Justice Samuel Alito (1950–) noted that "incorporation does not imperil every law regulating firearms," and Chicago and other municipalities throughout the country immediately began revising their gun control policies.

West Hollywood, California: Saturday Night Specials Banned

In 1998 the California Supreme Court let stand a ruling by the California Court of Appeals on inexpensive handguns known as Saturday Night Specials. The court of appeals had upheld a municipal ban by the city of West Hollywood on the sale of this type of weapon. In its opinion, the court rejected the gun lobby's claim that California state law preempted the ordinance; furthermore, the court found that the ban did not violate the principles of equal protection or due process. Gun rights advocates maintained that singling out inexpensive weapons denies poor people an affordable means of self-defense.

Denver, Colorado: State Law Does Not Override Denver's Right to Ban Certain Guns

In 2003 the Colorado state legislature passed gun legislation that preempted many of Denver's local firearms laws. The city of Denver filed a lawsuit against the state of Colorado and Governor Bill Owens (1950–) to retain its city ordinances. Judges Joseph E. Meyer III and Lawrence Manzanares (1956–2007) of the Denver District Court ruled that the state legislation did and could override some of Denver's minor ordinances, but that the city still had the right to ban certain guns such as assault weapons and some handguns.

Seattle, Washington: State Law Overrides City Gun Ban

In October 2009 the Second Amendment Foundation and several other gun rights groups sued the city of Seattle, claiming that the city had no right to ban guns in certain places, such as parks and community centers. Judge Catherine Shaffer of the King County Superior Court ruled in February 2010 that Washington State law did not allow Seattle to regulate the possession of firearms. The city of Seattle appealed the ruling to the U.S. District Court on Second Amendment grounds, citing *Heller.* In March 2010 Judge Marsha J. Pechman (1951–) of the U.S. District Court for the Western District of Washington upheld the lower court ruling and stated that the *Heller* ruling applied only to federal jurisdictions. However, the U.S. Supreme Court ruling in *McDonald v. Chicago* in June 2010 clarified that Second

Amendment rights also apply to the states. The article "Justices Extend Gun Owner Rights Nationwide; McKenna Statement Inside" (KHQ.com, June 30, 2010) quotes Rob McKenna (1962–), the attorney general of Washington State, as saying that he was "gratified that the U.S. Supreme Court has affirmed that the Second Amendment may not be infringed by state and local governments" and that "the right to bear arms shouldn't be infringed just because a person crosses state or city lines."

RESPONSIBILITY FOR HANDGUN DEATHS

The cases presented thus far on the federal, state, and local levels focus primarily on the right to bear arms. The following cases probe the responsibilities and liabilities associated with the use of those arms. Some victims of the use of certain weapons have tried to place that responsibility and liability on the manufacturers of the weapons, whereas others fault the people who made the weapons available to criminals. Each of the decisions presented here was based on state laws that differ greatly.

California: Gun Dealers Can Be Held Liable for Gun Violations of Others

Nineteen-year-old Jeff Randa had mentioned to a gun dealer many times that he wanted to buy a handgun and ammunition. The dealer told the youth that he could not buy a gun until he was 21 years old. Randa asked if his grandmother could purchase the weapon. The dealer replied that if she were a qualified buyer she could, but that the dealer could not sell her the weapon "just so she could give the gun to her grandson."

Subsequently, Randa's grandmother came into the store with him and purchased the handgun the youth wanted. Twelve days later, Randa took the gun to a party. Bryan Hoosier, who was also at the party, told Randa to point the gun and shoot. Randa did so, killing Hoosier, and was later convicted of voluntary manslaughter.

Hoosier's father sued the gun dealer for negligence, accusing the dealer of knowing that the gun would be given to a minor after being sold to an adult. The dealer argued that he could not be liable and that the state laws imposed criminal penalties only on violators.

The California Court of Appeals ruled in *Hoosier v. Randa* (17 Cal. Rptr. 2d 518, 521 [Cal. Ct. App. 1993]) that the dealer was indeed liable for injuries. The state gun control laws were passed not only to establish criminal penalties but also to protect the public. If a dealer violated the law, he also violated his responsibility of care owed to the public. Consequently, any person harmed by such a violation may sue the violator.

Ohio: Gun Show Promoters Must Provide Adequate Security against Juvenile Gun Theft

During a 1992 gun show that was promoted by Niles Gun Show Inc., four youths under the age of 18 years stole several handguns. The corporation from which the vendors rented space had no policy that required the vendors to protect their wares from being stolen, although it had an unenforced policy barring minors from entering the show.

After leaving the show, the youths also stole a car. While driving around in the car, the juveniles confronted two men, Greg L. Pavlides and Thomas E. Snedeker. One of the boys, Edward A. Tilley III, shot Pavlides in the chest and Snedeker in the head with one of the stolen guns. Tilley was arrested, charged, and convicted of two counts of attempted murder and one count of unauthorized use of a motor vehicle.

Pavlides and Snedeker survived their injuries and sued Niles Gun Show for negligence for not protecting them and the rest of the public from criminal acts by third parties who stole weapons that had not been properly secured. The trial court dismissed the case, stating that the promoters had no such responsibility, but the Ohio Court of Appeals reversed the lower court's decision, sending the case back to be tried. The court ruled that the promoters of gun shows have a duty to provide adequate security to protect the public from criminal acts that might occur if guns are stolen. The court explained that gun show operators should require vendors to secure their firearms and make a reasonable effort to bar minors from stealing or purchasing weapons and ammunition. The court further stated that it is "common knowledge" that minors possessing guns can create dangerous situations, and consequently gun show promoters should be aware that minors stealing guns might use them in criminal activity.

Texas: Gun Seller Not Liable for Purchaser's Suicide

In December 1980 James J. Robertson purchased a handgun. Eighteen months later he used that gun to kill himself. Robertson's family brought a wrongful-death suit against the seller of the handgun before the 298th Judicial District Court, Dallas County, Texas. In *Robertson v. Grogan Investment Company* (710 SW.2d 678 [1986]), the district court found in favor of the defendant, Grogan Investments, because "the sale of handguns ... to the general public is an abnormally dangerous and ultrahazardous activity.... Texas courts, when confronted with the opportunity to apply strict liability for ultrahazardous activities, have declined to do so and have consistently required some other showing, such as negligence or trespass, for recovery."

Florida: Store Responsible in Criminal Act for Selling Ammunition to Juveniles

In 1991 a Florida Wal-Mart store employee sold ammunition to two teenagers without asking about age or requesting identification, which is a violation of federal law. Several hours later the teenagers used the ammunition in a robbery of an auto parts store, during which they fatally shot Billy Wayne Coker. Coker's wife filed suit against Wal-Mart.

Although Wal-Mart acknowledged that the sale was illegal, it argued that the perpetrators' intervening act of murder was not foreseeable and, therefore, the illegal sale was not the legal cause of Coker's death. The court agreed with this argument and dismissed the case. However, the Florida Court of Appeals ruled that an ammunition vendor's illegal sale could be the legal cause of an injury or death caused by the buyer's intentional or criminal act. In July 1998 the Florida Supreme Court upheld in *Wal-Mart Stores, Inc. v. Coker* (1998 Fla. Lexis 861) the $2.6 million verdict against Wal-Mart for negligence in selling handgun ammunition to underaged buyers.

GUN INDUSTRY AND LIABILITY FOR GUNSHOT INJURIES
Maryland: *Kelley v. R.G. Industries Inc.*

In 1981 Olen J. Kelley was injured when he was shot in the chest during an armed robbery of the grocery store where he worked. The gun used was a Rohm revolver handgun model RG-38S that was designed and marketed by Rohm Gesellschaft, a German corporation. The handgun was assembled and initially sold by R.G. Industries Inc., a Miami-based subsidiary of the German corporation. Kelley and his wife filed suits against Rohm Gesellschaft and R.G. Industries in the Circuit Court of Montgomery County, Maryland.

Two counts charged that the handgun was "abnormally dangerous" and "defective in its marketing, promotion, distribution, and design." A third count charged negligence. The case revolved around whether the gun in question was a Saturday Night Special, which was banned from import by the ATF. The Federal District Court of Baltimore, where the case was first brought, asked the state court for a ruling on whether the manufacturer could be held liable under Maryland law.

The Maryland Court of Appeals ruled in *Kelley v. R.G. Industries Inc.* (497 A.2d 1143 [1985]) that the manufacturer and marketers could not be held strictly liable because handguns are "abnormally dangerous products" and their manufacturing and marketing are "abnormally dangerous activit[ies]." In its decision, the court noted, "Contrary to Kelley's argument, a handgun is not defective merely because it is capable of being used during criminal activity to inflict harm. A consumer would expect a handgun to be dangerous, by its very nature, and to have the capacity to fire a bullet with deadly force."

The court of appeals also stated that Kelley confused a product's normal function, which may be dangerous by its very nature, with a defect in its design and function. Kelley had cited as an example that a car is dangerous if it is used to run down pedestrians. The injury that results is from the nature of the product—the ability to be propelled to great speeds at great force. Nevertheless, if the gas tank of the car leaked in such a way as to cause an explosion in the event of a rear-end collision, then the design of the product would be defective, and the manufacturer would be liable. The court concluded that to impose "strict liability upon the manufacturers or marketers of handguns for gunshot injuries resulting from the misuse of handguns by others, would be contrary to Maryland public policy."

The Maryland court's opinion differed on Saturday Night Specials, which it defined as guns "characterized by short barrels, light weight, easy concealability, low cost, use of cheap quality materials, poor manufacture, inaccuracy and unreliability." The court considered these guns "largely unfit for any of the recognized legitimate uses sanctioned by the Maryland gun control legislation. They are too inaccurate, unreliable and poorly made for use by law enforcement personnel, sportsmen, homeowners or businessmen.... The chief 'value' a Saturday Night Special handgun has is in criminal activity, because of its easy concealability and low price."

The court determined that manufacturers and marketers are liable because they should know that this type of gun is made primarily for criminal activity. Judge John C. Eldridge (1933–) quoted an R.G. Industries salesperson as telling a prospective handgun marketer, "If your store is anywhere near a ghetto area, these ought to sell real well. This is most assuredly a ghetto gun." The salesperson allegedly went on to say that although the gun sold well, it was virtually useless, and that he would be afraid to fire it.

The court of appeals did not rule on whether the gun in question fell within the category of Saturday Night Specials but referred that decision to the U.S. District Court. The court of appeals, however, did indicate that strong evidence had been presented that the gun fit many of the qualifications; if it were found to be a Saturday Night Special, liability against both the manufacturer and marketer could be imposed. This decision applied only in Maryland, and the Maryland legislature soon passed a law overriding it. Few courts have accepted this interpretation.

New Mexico: *Armijo v. Ex Cam Inc.*

Dolores Armijo's brother, Steven Armijo, shot and killed James Salusberry, Dolores's husband, in front of Dolores and her daughter. He then tried to shoot them, but the gun jammed. Dolores, claiming the gun used was a Saturday Night Special, sued Ex Cam Inc., the importer and distributor of the weapon.

The suit was based on four theories: strict product liability (the product was defective and unreasonably dangerous; therefore, the manufacturer was responsible for the actions of the product), "ultra-hazardous activity" liability (a gun is a dangerous product and the manufacturer is accountable for the results of its use), negligence liability (the manufacturer did not show reasonable care while marketing a product that carried some degree of risk that it might be used to commit a crime), and a narrow form of strict product liability for Saturday Night Specials put forth in *Kelley v. R.G. Industries Inc.*

The U.S. District Court in New Mexico did not believe that any court in New Mexico would ever recognize any of these theories as the basis of a court case under New Mexico law. In *Armijo v. Ex Cam Inc.* (656 F.Supp. 771 [1987]), the court said:

> It would be evident to any potential consumer that a gun could be used as a murder weapon. So could a knife, an axe, a bow and arrows, a length of chain. The mere fact that a product is capable of being misused to criminal ends does not render the product defective.

> [Based on New Mexico law, such a case] would not result in liability for a manufacturer of guns, as guns are commonly distributed and the dangers ... are so obvious as to not require any manufacturers' warnings.

The court showed little respect for *Kelley v. R.G. Industries Inc.*, indicating that it went against common law in the state of New Mexico; therefore, it would not be considered. Furthermore, the court concluded that "all firearms are capable of being used for criminal activity. Merely to impose liability upon the manufacturers of the cheapest types of handguns will not avoid that basic fact. Instead, claims against gun manufacturers will have the anomalous [unusual] result that only persons shot with cheap guns will be able to recover, while those shot with expensive guns, admitted by the *Kelley* court to be more accurate and therefore deadlier, would take nothing."

Washington, D.C.: *Delahanty v. Hinckley*

In 1981 John W. Hinckley Jr. (1955–) tried to assassinate President Ronald Reagan (1911–2004). An individual injured during the assassination attempt sued to hold the gun manufacturer liable based on negligence, strict product liability, and a "social utility" claim founded on strict liability for unusually dangerous products.

In *Delahanty v. Hinckley* (DC, No. 88-488 [1989]), the Washington, D.C., court rejected the plaintiff's claims. There was no issue that the gun did not work properly. Furthermore, a manufacturer had no duty to warn a buyer "when the danger, or potentiality of danger, is generally known and recognized."

The court did not believe the marketing of a handgun was in and of itself dangerous; rather, the danger resulted from the action of a third party. The plaintiff had shown no connection between the gun manufacturer and Hinckley, nor had the plaintiff shown a reasonable way in which the gun manufacturer could have prevented Hinckley from using the weapon to try to assassinate President Reagan. The court also dismissed the *Kelley* argument because it could not accept a ruling that categorizes one type of product as liable for negligence simply because it is inexpensive and/or poorly made.

Washington, D.C.: The Federal Government Settles the Manufacturer Liability Issue

In 1998, after the tobacco industry was found to be responsible for lung cancer deaths caused by smoking cigarettes, many cities and counties across the United States began filing lawsuits against gun manufacturers and dealers. At issue was the gun distribution system, which was thought to allow guns to pass too easily to criminals and youth. However, Fox Butterfield reports in "Gun Industry Is Gaining Immunity from Suits" (NY Times.com, September 1, 2002) that by 2002, 30 states had passed laws granting immunity to the gun industry from these civil lawsuits. In a countermeasure, California passed a bill repealing such an immunity law in its state.

In 2005 Congress settled this issue with the passage of the Protection of Lawful Commerce in Arms Act. President George W. Bush (1946–) signed the bill into law in October 2005. This act prohibits liability actions (charging legal responsibility) against firearms and ammunition manufacturers and sellers for unlawful misuse of their products. An exception to the act allows petitioners to sue firearms manufacturers and dealers if they knowingly violate state or federal statutes. The act also prohibits the sale of a handgun unless the purchaser is provided with a secure gun storage or safety device, and it provides for particular sentences when armor-piercing ammunition is used in certain crimes.

According to David Stout, in "Justices Decline New York Gun Suit" (NYTimes.com, March 9, 2009), since the passage of the Protection of Lawful Commerce in Arms Act, many city and state officials have sued gun

manufacturers without success, but some litigants have won suits against gun dealers.

Williams v. Beemiller: Gun Industry May Be Liable When Engaged in Illegal Trafficking

In *Williams v. Beemiller, Inc.* (100 A.D.3d 143 [2012]), the New York Court of Appeals, Fourth Judicial Department, reversed a lower court's dismissal of a case in which a shooting victim brought suit against the gun manufacturer, distributor, and dealer who supplied the weapon used to injure him. *Williams v. Beemiller* stemmed from a drive-by shooting that occurred in Buffalo, New York, in 2003. In a case of mistaken identity, teenager Daniel Williams was shot by gang member Cornell Caldwell with a 9mm Hi-Point handgun that had been acquired by James Nigel Bostic in a straw purchase, with his girlfriend as the buyer of record. Bostic had purchased about 250 guns in Ohio that he sold illegally in Buffalo; 140 of the guns were purchased from one dealer, Charles Brown, and 87 guns were exchanged in the transaction that included the one that was used to shoot Williams. Caldwell was apprehended, tried, found guilty, and sent to prison for the crime.

Williams, who survived the shooting, filed a suit against the manufacturer of the gun, the distributor, and Brown, who he claimed should have recognized that Bostic was involved in criminal activity from the number and type of weapons that he purchased. Brown claimed that Bostic told him he was planning to open a gun shop. In 2011 the defendants had sought and received a dismissal of the case based on the Protection of Lawful Commerce in Arms Act; however, the New York State Court of Appeals overturned that dismissal in October 2012. Justice Erin Peradotto of the Appellate Division, Fourth Department, stated in the ruling, "Although the complaint does not specify the statutes allegedly violated (by the defendants), it sufficiently alleges facts supporting a finding that defendants knowingly violated federal gun laws." The ruling allowed Williams's case to move forward, therefore, holding open the possibility that the gun industry may be criminally liable when engaged in illegal sales. As of November 2014, this case had not been resolved

CHAPTER 5
FIREARMS AND CRIME

Many of the statistics on the frequency and ways in which guns are used to commit crimes come from the Federal Bureau of Investigation (FBI) of the U.S. Department of Justice (DOJ), which collects a range of U.S. crime statistics. One of the primary sources of FBI statistics is the Uniform Crime Reports (UCR) program, on which the annual publication *Crime in the United States* is based. *Crime in the United States* and other UCR data releases are derived from police investigation reports and arrests; crimes that are not reported to the police are not included. As of November 2014, the most recent final report from the UCR program was *Crime in the United States, 2013* (2013, http://www.fbi.gov/about-us/cjis/ucr/crime-in-the-u.s/2013/crime-in-the-u.s.-2013/cius-home). The FBI's Bureau of Justice Statistics (BJS) also regularly releases publications based on its National Crime Victimization Survey (NCVS). The NCVS, which consists of a survey of a representative sample of U.S. households, seeks to measure rates and levels of victimization for nonfatal violent and property crimes. It is different from the UCR program into two important ways. First, it attempts to capture both crimes that are reported to police and those that are not. Second, it cannot capture information about homicides because it consists of surveys of living people. As such, UCR and NCVS statistics are not identical, but they are often broadly consistent with each other. They complement one another in key ways and thus contribute to a comprehensive, albeit incomplete, understanding of criminal activity and incidents in the United States.

In *Crime in the United States, 2013*, the FBI reports that an estimated 1,163,146 violent crimes were reported to law enforcement agencies in 2013, which was a slight decrease from the previous year. The five-year period between 2009 and 2013 saw violent crimes reported to law enforcement fall significantly. (See Figure 5.1.) The 2013 level of violent crime was 12.3% lower than in

2009 and 14.5% lower than in 2004, and the overall rate of violent crime in the United States was 367.9 per 100,000 residents.

The FBI divides violent crime into four categories: murder and nonnegligent manslaughter, forcible rape, robbery, and aggravated assault. The category of murder and nonnegligent manslaughter is defined as "the willful (nonnegligent) killing of one human being by another." Prior to 2013 forcible rape was defined as "the carnal knowledge of a female forcibly and against her will." However, in 2013 the UCR program redefined this term by dropping the word *forcible*. Rape is now defined as "penetration, no matter how slight, of the vagina or anus with any body part or object, or oral penetration by a sex organ of another person, without the consent of the victim," including "attempts or assaults to commit rape" but excluding "statutory rape and incest." Although some UCR statistics were reported under this new definition of rape, others were reported under the older definition. The 2013 data reflect all reported rapes but are categorized according to the older definition to facilitate comparison with previous years. Robbery is defined as "the taking or attempting to take anything of value from the care, custody, or control of a person or persons by force or threat of force or violence and/or by putting the victim in fear." Aggravated assault is defined as "an unlawful attack by one person upon another for the purpose of inflicting severe or aggravated bodily injury." The FBI includes in this category, "Attempted aggravated assault that involves the display of—or threat to use—a gun, knife, or other weapon," and if aggravated assault occurs in combination with larceny-theft (a property crime), the bureau classifies the offense as robbery. Together, these four categories of crime represent those in which firearms are typically used. Firearms were used in 69% of U.S. murders, 40% of robberies, and 21.6% of aggravated assaults. Forcible rape reports to law enforcement are not classified according to weapons used.

FIGURE 5.1

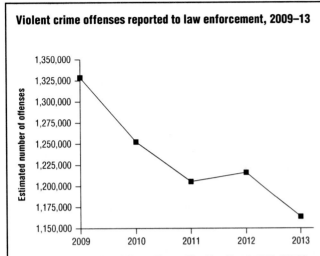

Violent crime offenses reported to law enforcement, 2009–13

SOURCE: "Violent Crime Offense Figure: Five Year Trend, 2009–2013," in "Violent Crime," *Crime in the United States, 2013*, Federal Bureau of Investigation, 2014, http://www.fbi.gov/about-us/cjis/ucr/crime-in-the-u.s/2013/crime-in-the-u.s.-2013/violent-crime/violent-crime-topic-page/violentcrimemain_final (accessed November 20, 2014)

In assessing the distribution of violent crime in the United States in 2013, the FBI divides cities into six groups: Group I, cities with populations of 250,000 and over, which had a total 2013 population of 57.4 million; Group II, cities with populations of 100,000 to 249,999, which had a total population of 32 million; Group III, cities with populations of 50,000 to 99,999, which had a total population of 32.5 million; Group IV, cities with populations of 25,000 to 49,999, which had a total population of 29 million; Group V, cities with populations of 10,000 to 24,999, which had a total population of 28.9 million; and Group VI, cities with populations under 10,000, which had a total population of 23.2 million. (See Table 5.1.) Group I cities, which made up approximately 19% of the total U.S. population, accounted for 37% of the total number of violent crimes, including 5,356 murders (39% of all U.S. murders), 15,522 forcible rapes (25%), 162,815 robberies (49%), and 219,295 aggravated assaults (32%).

HOMICIDE AND MURDER

Murders, Weapons, and Circumstances

In *Homicide Trends in the United States, 1980–2008* (November 2011, http://www.bjs.gov/content/pub/pdf/htus8008.pdf), Alexia Cooper and Erica L. Smith of the BJS note that the homicide victimization rate (the number of people killed by another per 100,000 population) was relatively low during the 1950s and the early 1960s, under five homicides per 100,000 people in most years. By the mid-1960s, however, the rate rose dramatically and continued to rise through the mid-1970s, reaching nearly 10 homicides per 100,000 people. Following a

slight dip, it peaked in 1980 at 10.2 homicides per 100,000 people. After a decline to 7.9 homicides per 100,000 people in 1984, the homicide rate rose again during the late 1980s and early 1990s to 9.8 homicides per 100,000 people in 1991. The homicide rate fell throughout the 1990s. In 2000 it reached the lowest level since 1966, at 5.5 homicides per 100,000 people, and remained stable through 2008. Smith and Cooper indicate in *Homicide in the U.S. Known to Law Enforcement, 2011* (December 2013, http://www.bjs.gov/content/pub/pdf/hus11.pdf) that after 2008 the homicide rate declined further, reaching 4.7 per 100,000 people in 2011, the lowest rate since 1963 and one that represented a 54% decline from the 1980 peak.

DEMOGRAPHICS OF HOMICIDE. The vast majority of murderers and murder victims are men. Smith and Cooper note that between 1992 and 2011 the homicide rate for males and females declined almost equally (by 50% for males and 49% for females), but over that whole period the homicide rate for males remained, on average, 3.6 times higher than the rate for females. In 2011 the homicide rate for males was 7.4 per 100,000, and there were 11,370 male murder victims. (See Table 5.2.) By comparison, the homicide rate for females was 2 per 100,000, and there were 3,240 female murder victims.

According to Smith and Cooper, over the 10-year period between 2002 and 2011 the homicide rate for African Americans was, on average, 6.3 times higher than the homicide rate for whites. During this period, however, homicides among both groups declined at similar rates (19% for African Americans and 17% for whites). In 2011 the homicide rate for African Americans was 17.3 per 100,000, compared with a rate of 2.8 per 100,000 for whites. (See Table 5.2.)

Between 2002 and 2011 the homicide victimization rate for both African American and white males peaked in the years of early adulthood, at age 23 for African Americans and at age 20 for whites. (See Figure 5.2.) However, Smith and Cooper indicate that the highest homicide rate for African American males, at 100.3 per 100,000 for males aged 23 years, was nine times greater than the highest homicide rate for white males, at 11.4 per 100,000 for males aged 20 years. Although the likelihood of being a murder victim declined with age for both groups, it remained significantly higher for African American males at all ages. For example, at age 60 African American men were over four times more likely to be homicide victims than white men.

A similar pattern applied to homicide victimization rates for African American and white women between 2002 and 2011, but the overall rates were much lower, and there were key differences among men and women regarding peak ages for homicide victimization. (See Figure 5.3.) Smith and Cooper explain that both African

TABLE 5.1

Crime trends, by population group, 2012–13

Population group		Violent crime	Murder and nonnegligent manslaughter	Forcible rape[a]	Robbery	Aggravated assault	Property crime	Burglary	Larceny-theft	Motor vehicle theft	Arson	Number of agencies	2013 estimated population
Total all agencies:	2012	1,145,272	14,349	65,733	345,758	719,432	8,507,866	1,992,895	5,816,991	697,980	51,126	15,232	299,269,511
	2013	1,095,149	13,716	62,034	335,428	683,971	8,160,228	1,820,544	5,665,392	674,292	44,245		
	Percent change	−4.4	−4.4	−5.6	−3	−4.9	−4.1	−8.6	−2.6	−3.4	−13.5		
Total cities	2012	919,218	11,198	50,003	303,475	554,542	6,671,809	1,460,169	4,655,398	556,242	38,974	11,031	202,966,923
	2013	877,594	10,511	47,606	294,292	525,185	6,434,879	1,335,527	4,557,867	541,485	33,432		
	Percent change	−4.5	−6.1	−4.8	−3	−5.3	−3.6	−8.5	−2.1	−2.7	−14.2		
Group I (250,000 and over)	2012	416,885	5,897	15,715	167,168	228,105	2,155,161	493,768	1,401,823	259,570	12,945	77	57,394,814
	2013	402,988	5,356	15,522	162,815	219,295	2,097,875	454,980	1,395,583	247,312	12,150		
	Percent change	−3.3	−9.2	−1.2	−2.6	−3.9	−2.7	−7.9	−0.4	−4.7	−6.1		
1,000,000 and over (Group I subset)	2012	166,007	2,255	5,254	74,843	83,655	783,251	166,197	523,397	93,657	3,547	10	25,735,804
	2013	161,252	1,930	5,356	71,478	82,488	760,143	151,503	521,708	86,932	3,612		
	Percent change	−2.9	−14.4	1.9	−4.5	−1.4	−3	−8.8	−0.3	−7.2	1.8		
500,000 to 999,999 (Group I subset)	2012	138,439	1,929	5,833	48,874	81,803	747,008	177,258	481,094	88,656	4,772	24	16,780,477
	2013	134,216	1,786	5,848	48,375	78,207	733,283	164,283	482,351	86,649	4,466		
	Percent change	−3.1	−7.4	0.3	−1	−4.4	−1.8	−7.3	0.3	−2.3	−6.4		
250,000 to 499,999 (Group I subset)	2012	112,439	1,713	4,628	43,451	62,647	624,902	150,313	397,332	77,257	4,626	43	14,878,533
	2013	107,520	1,640	4,318	42,962	58,600	604,449	139,194	391,524	73,731	4,072		
	Percent change	−4.4	−4.3	−6.7	−1.1	−6.5	−3.3	−7.4	−1.5	−4.6	−12		
Group II (100,000 to 249,999)	2012	154,144	1,788	8,812	50,456	93,088	1,167,627	268,959	793,349	105,319	6,540	215	32,029,761
	2013	145,408	1,806	7,951	48,590	87,061	1,130,942	246,360	778,577	106,005	5,603		
	Percent change	−5.7	1	−9.8	−3.7	−6.5	−3.1	−8.4	−1.9	0.7	−14.3		
Group III (50,000 to 99,999)	2012	116,180	1,162	7,351	34,531	73,136	977,397	215,072	687,344	74,981	5,584	471	32,516,218
	2013	110,033	1,171	6,972	33,264	68,626	942,519	194,902	673,912	73,705	4,599		
	Percent change	−5.3	0.8	−5.2	−3.7	−6.2	−3.6	−9.4	−1.7	−1.7	−17.6		
Group IV (25,000 to 49,999)	2012	85,900	920	6,370	23,082	55,528	836,125	172,406	616,275	47,444	4,538	841	28,988,979
	2013	80,441	865	6,035	22,462	51,079	806,275	158,056	601,616	46,603	3,489		
	Percent change	−6.4	−6	−5.3	−2.7	−8	−3.6	−8.3	−2.4	−1.8	−23.1		
Group V (10,000 to 24,999)	2012	77,759	826	6,011	17,517	53,405	817,630	169,750	608,579	39,301	3,809	1,813	28,865,002
	2013	73,998	713	5,587	17,065	50,633	778,955	153,632	586,929	38,394	3,212		
	Percent change	−4.8	−13.7	−7.1	−2.6	−5.2	−4.7	−9.5	−3.6	−2.3	−15.7		
Group VI (under 10,000)	2012	68,350	605	5,744	10,721	51,280	717,869	140,214	548,028	29,627	5,558	7,614	23,172,149
	2013	64,726	600	5,539	10,096	48,491	678,313	127,597	521,250	29,466	4,379		
	Percent change	−5.3	−0.8	−3.6	−5.8	−5.4	−5.5	−9	−4.9	−0.5	−21.2		
Metropolitan counties[b]	2012	183,662	2,337	11,357	39,349	130,619	1,461,686	405,229	938,019	118,438	9,151	1,844	71,774,379
	2013	177,318	2,386	10,425	38,507	126,000	1,380,837	369,194	901,047	110,596	8,153		
	Percent change	−3.5	2.1	−8.2	−2.1	−3.5	−5.5	−8.9	−3.9	−6.6	−10.9		
Nonmetropolitan counties[b]	2012	42,392	814	4,373	2,934	34,271	374,371	127,497	223,574	23,300	3,001	2,357	24,528,209
	2013	40,237	819	4,003	2,629	32,786	344,512	115,823	206,478	22,211	2,660		
	Percent change	−5.1	0.6	−8.5	−10.4	−4.3	−8	−9.2	−7.6	−4.7	−11.4		
Suburban areas[c]	2012	326,107	3,773	21,495	75,543	225,296	3,023,845	709,426	2,112,377	202,042	17,202	8,346	131,056,632
	2013	312,046	3,692	20,220	73,494	214,640	2,865,297	642,401	2,030,729	192,167	14,556		
	Percent change	−4.3	−2.1	−5.9	−2.7	−4.7	−5.2	−9.4	−3.9	−4.9	−15.4		

[a]The rape figures in this table are based on the legacy Uniform Crime Reporting (UCR) definition of rape. The rape figures shown for 2012 and 2013 include converted National Incident-Based Reporting System rape data and those states/agencies that reported the legacy UCR definition of rape for both years.

[b]Includes state police agencies that report aggregately for the entire state.

[c]Suburban areas include law enforcement agencies in cities with less than 50,000 inhabitants and county law enforcement agencies that are within a Metropolitan Statistical Area. Suburban areas exclude all metropolitan agencies associated with a principal city. The agencies associated with suburban areas also appear in other groups within this table.

SOURCE: "Table 12. Crime Trends by Population Group, 2012–2013," in *Crime in the United States, 2013*, Federal Bureau of Investigation, 2014, http://www.fbi.gov/about-us/cjis/ucr/crime-in-the-u.s/2013/crime-in-the-u.s.-2013/tables/table-12/table_12_crime_trends_by_population_group_2012-2013.xls (accessed November 20, 2014)

TABLE 5.2

Number and rate of homicides, by victim demographic characteristics, 2002–11

Year	All victims	Sex		Race			Age						
		Male	Female	White	Black/African American	Other*	11 or younger	12–17	18–24	25–34	35–49	50–64	65 or older
Number of homicides													
2002	16,230	12,475	3,755	7,805	7,990	435	790	825	4,345	4,305	3,875	1,365	720
2003	16,530	12,835	3,690	7,985	8,080	465	790	785	4,510	4,375	3,890	1,425	750
2004	16,150	12,600	3,550	7,980	7,755	415	765	860	4,165	4,295	3,800	1,510	755
2005	16,740	13,180	3,560	8,050	8,255	435	745	925	4,405	4,480	3,880	1,560	740
2006	17,310	13,655	3,655	8,135	8,710	460	780	1,035	4,615	4,545	3,920	1,755	660
2007	17,130	13,460	3,665	8,110	8,610	405	800	1,035	4,485	4,570	3,875	1,680	675
2008	16,465	12,900	3,565	8,020	8,070	375	810	955	4,065	4,445	3,745	1,705	745
2009	15,400	11,880	3,520	7,485	7,495	425	730	840	3,840	3,975	3,590	1,685	745
2010	14,720	11,410	3,315	6,885	7,450	385	710	770	3,655	4,025	3,200	1,700	665
2011	14,610	11,370	3,240	6,830	7,380	400	740	665	3,680	3,850	3,260	1,710	700
Rate per 100,000 U.S. residents													
2002	5.6	8.8	2.6	3.3	21.2	2.7	1.6	3.3	15.2	10.9	5.9	3.0	2.0
2003	5.7	9.0	2.5	3.4	21.2	2.9	1.6	3.1	15.6	11.1	5.9	3.0	2.1
2004	5.5	8.7	2.4	3.4	20.1	2.5	1.6	3.4	14.2	10.9	5.8	3.1	2.1
2005	5.7	9.1	2.4	3.4	21.2	2.5	1.5	3.6	14.9	11.3	5.9	3.1	2.0
2006	5.8	9.3	2.4	3.4	22.1	2.6	1.6	4.1	15.5	11.4	6.0	3.4	1.8
2007	5.7	9.1	2.4	3.3	21.5	2.2	1.6	4.1	15.0	11.4	5.9	3.1	1.8
2008	5.4	8.6	2.3	3.3	19.9	2.0	1.6	3.8	13.4	10.9	5.8	3.1	1.9
2009	5.0	7.8	2.3	3.0	18.3	2.2	1.5	3.4	12.5	9.6	5.6	3.0	1.9
2010	4.8	7.5	2.1	2.8	17.7	1.8	1.4	3.0	11.9	9.8	5.0	2.9	1.6
2011	4.7	7.4	2.0	2.8	17.3	1.8	1.5	2.7	11.9	9.2	5.2	2.8	1.7

*Includes persons identified as American Indian, Alaska Native, Asian, Native Hawaiian, or other Pacific Islander.
Note: Data may not sum to total due to rounding. Counts rounded to the nearest 5. Homicide rates by Hispanic or Latino origin were not calculated due to missing data on ethnicity.

SOURCE: Erica L. Smith and Alexia Cooper, "Table 1. Number and Rate of Homicides in the U.S., by Victim Demographic Characteristics, 2002–2011," in *Homicide in the U.S. Known to Law Enforcement, 2011*, U.S. Department of Justice, Bureau of Justice Statistics, December 2013, http://www.bjs.gov/content/pub/pdf/hus11.pdf (accessed July 15, 2014)

FIGURE 5.2

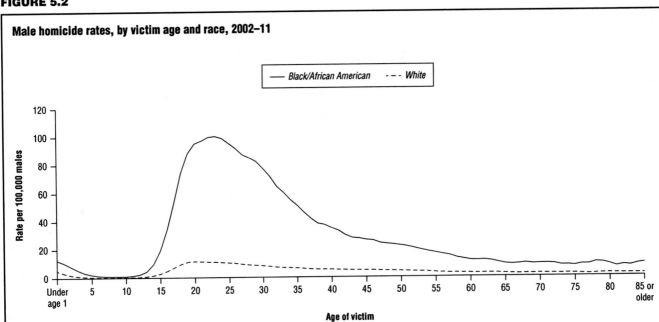

Male homicide rates, by victim age and race, 2002–11

Note: Homicide rates by Hispanic or Latino origin were not calculated due to missing data on ethnicity.

SOURCE: Erica L. Smith and Alexia Cooper, "Figure 5. Male Homicide Rates, by Victim Age and Race, 2002–2011," in *Homicide in the U.S. Known to Law Enforcement, 2011*, U.S. Department of Justice, Bureau of Justice Statistics, December 2013, http://www.bjs.gov/content/pub/pdf/hus11.pdf (accessed July 15, 2014)

FIGURE 5.3

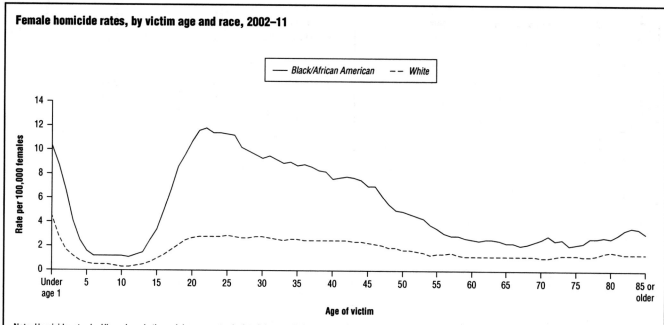

Female homicide rates, by victim age and race, 2002–11

Note: Homicide rates by Hispanic or Latino origin were not calculated due to missing data on ethnicity.

SOURCE: Erica L. Smith and Alexia Cooper, "Figure 6. Female Homicide Rates, by Victim Age and Race, 2002–2011," in *Homicide in the U.S. Known to Law Enforcement, 2011*, U.S. Department of Justice, Bureau of Justice Statistics, December 2013, http://www.bjs.gov/content/pub/pdf/hus11.pdf (accessed July 15, 2014)

American and white females had high homicide victimization rates during their first year of life. For white females, the homicide rate peaked at age one or under, at 4.5 per 100,000, before rising again during the teenage years and remaining relatively steady through middle age. For African American females the elevated homicide rate during the first year of life, at 10.3 per 100,000, was eclipsed by the rate at age 22, which was 11.8 per 100,000. During the young adult years African American females were roughly four times more likely than white females to be the victims of homicide, and between the ages of 30 and 50 years African American women remained over three times more likely than white women to be the victims of murder.

FIREARMS USED IN HOMICIDES. As noted earlier, firearms are the most common weapon used to commit murder. Between 1992 and 2011 the firearm homicide rate declined by almost half, from 6.3 per 100,000 people to 3.2 per 100,000. (See Figure 5.4.) However, the rate at which homicides were committed using firearms remained, as of 2011, three times higher than the rate at which homicides were committed using a knife or blunt object, which was the next most common weapon. Furthermore, Smith and Cooper indicate that the share of firearm homicides among all homicides remained steady over these two decades, at 73% for male homicide victims and 49% for female homicide victims. As Figure 5.5 and Figure 5.6 show, handguns were the most common type of firearm used to commit murder, and the percentage of murders committed with handguns remained

FIGURE 5.4

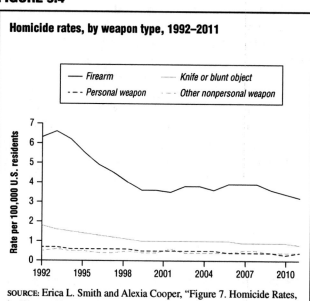

Homicide rates, by weapon type, 1992–2011

SOURCE: Erica L. Smith and Alexia Cooper, "Figure 7. Homicide Rates, by Weapon Type, 1992–2011," in *Homicide in the U.S. Known to Law Enforcement, 2011*, U.S. Department of Justice, Bureau of Justice Statistics, December 2013, http://www.bjs.gov/content/pub/pdf/hus11.pdf (accessed July 15, 2014)

relatively steady between 1992 and 2011. Over the course of these two decades an average of 57% of male homicides and 35% of female homicides were committed with handguns, while an average of 16% of male homicides and 13% of female homicides were committed with other types of firearms, such as rifles, shotguns, or assault weapons.

FIGURE 5.5

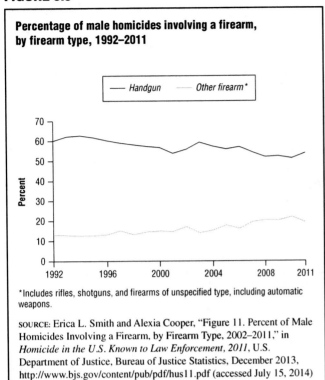

Percentage of male homicides involving a firearm, by firearm type, 1992–2011

[Legend: Handgun — Other firearm*]

*Includes rifles, shotguns, and firearms of unspecified type, including automatic weapons.

SOURCE: Erica L. Smith and Alexia Cooper, "Figure 11. Percent of Male Homicides Involving a Firearm, by Firearm Type, 2002–2011," in *Homicide in the U.S. Known to Law Enforcement, 2011*, U.S. Department of Justice, Bureau of Justice Statistics, December 2013, http://www.bjs.gov/content/pub/pdf/hus11.pdf (accessed July 15, 2014)

FIGURE 5.6

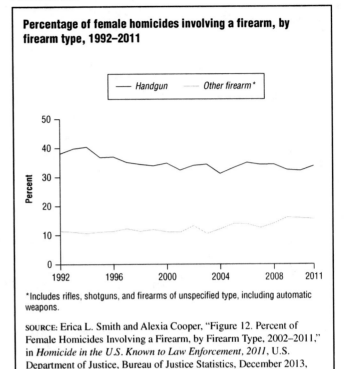

Percentage of female homicides involving a firearm, by firearm type, 1992–2011

[Legend: Handgun — Other firearm*]

*Includes rifles, shotguns, and firearms of unspecified type, including automatic weapons.

SOURCE: Erica L. Smith and Alexia Cooper, "Figure 12. Percent of Female Homicides Involving a Firearm, by Firearm Type, 2002–2011," in *Homicide in the U.S. Known to Law Enforcement, 2011*, U.S. Department of Justice, Bureau of Justice Statistics, December 2013, http://www.bjs.gov/content/pub/pdf/hus11.pdf (accessed July 15, 2014)

Table 5.3 shows murder circumstances by weapon in 2013. In that year firearms accounted for 8,454 of 12,253 (69%) murders. Handguns accounted for 5,782, or 47.2% of total murders, and for 68.4% of all firearm murders. Among the 1,909 murders committed in combination with another type of felony, 1,381 (71.9%) were committed with firearms, including 536 of 686 (78.1%) robberies and 327 of 386 (84.7%) narcotic drug offenses. Meanwhile, firearms accounted for 3,673 of 5,782 (63.5%) murders committed in circumstances that did not involve additional types of felonies, including 117 of 138 (84.8%) gangland killings and 547 of 584 (93.7%) juvenile gangland killings. Among murders committed as a result of arguments, including arguments over money or property (which resulted in 133 murders) and other arguments (2,889), firearms were the murder weapon in 60.6% of cases reported to law enforcement agencies.

Table 5.4 shows murder by state and type of weapon in 2013, among cases in which supplemental homicide data were available. As a result, the total number of murders for each state is not necessarily comprehensive but rather represents only those for which supplemental data about weapon type were available. Among such homicide cases, firearms were the most likely type of weapon used, and handguns were the most common type of firearm used. For example, firearms accounted for 1,224 of 1,745 (70.1%) murders in California and 760 of 1,133 (67.1%) murders in Texas, the states with the highest homicide totals. Similar or higher percentages

were evident in other states with high murder totals, including Georgia (where 411 of 534, or 77%, of murders were committed with firearms), Illinois (364 of 433, or 84.1%), Louisiana (356 of 453, or 78.6%), Maryland (268 of 379, or 70.7%), Michigan (440 of 625, or 70.4%), Missouri (273 of 371, or 73.6%), New Jersey (291 of 401, or 72.6%), New York (362 of 648, or 55.9%), North Carolina (315 of 452, or 69.7%), Ohio (309 of 434, or 71.2%), Pennsylvania (440 of 594, or 74.1%), and Tennessee (223 of 327, or 68.2%).

Murder often occurs at the hands of a family member, friend, romantic partner, or acquaintance. As Table 5.5 shows, the relationship of the victim to the murderer was unknown in 5,572 of 12,253 (45.5%) homicide cases. In 2,660 (21.7%) cases the victim and murderer were acquaintances. In 1,419 (11.6%) cases the victim was either the husband, wife, mother, father, son, daughter, brother, sister, or other family member of the perpetrator. Among all murders of family members, wives were victims in 534 cases, or 37.6% of the total. Friend (346 cases) and girlfriend (458) were among the other most common relationship type among murder victims.

As the previous data indicate, women who are murdered are especially likely to have a preexisting relationship with those who kill them. In *When Men Murder Women: An Analysis of 2012 Homicide Data* (September 2014, http://www.vpc.org/studies/wmmw2014.pdf), the Violence Policy Center, a national nonprofit educational

TABLE 5.3

Homicide circumstances, by weapon, 2013

Circumstances	Total murder victims	Totals	Handguns	Rifles	Shotguns	Other guns or type not stated	Knives or cutting instruments	Blunt objects (clubs, hammers, etc.)	Personal weapons (hands, fists, feet, etc.)	Poison	Pushed or thrown out window	Explosives	Fire	Narcotics	Drowning	Strangulation	Asphyxiation	Other
Total	**12,253**	**8,454**	**5,782**	**285**	**308**	**2,079**	**1,490**	**428**	**686**	**11**	**1**	**2**	**94**	**53**	**4**	**85**	**95**	**850**
Felony type total:	**1,909**	**1,381**	**1,026**	**32**	**36**	**287**	**174**	**79**	**64**	**1**	**1**	**0**	**42**	**21**	**0**	**17**	**11**	**118**
Rape*	20	2	2	0	0	0	3	4	6	0	0	0	0	0	0	1	1	3
Robbery	686	536	435	8	10	83	52	39	20	0	0	0	1	0	0	6	5	27
Burglary	94	52	29	2	2	19	24	5	5	0	0	0	2	1	0	0	1	5
Larceny-theft	16	6	4	0	0	2	3	2	2	0	0	0	2	0	0	0	0	5
Motor vehicle theft	27	12	8	0	2	2	4	2	0	0	0	0	2	0	0	0	0	7
Arson	37	1	0	0	0	1	6	1	1	0	0	0	22	0	0	0	0	6
Prostitution and commercialized vice	13	7	4	0	0	3	3	0	1	0	0	0	0	0	0	1	0	1
Other sex offenses	9	0	0	0	0	0	3	3	1	0	0	0	0	0	0	2	0	0
Narcotic drug laws	386	327	244	5	8	70	25	4	4	0	0	0	1	18	0	0	1	6
Gambling	7	6	5	0	0	1	1	0	0	0	0	0	0	0	0	0	0	0
Other-not specified	614	432	295	17	14	106	50	22	22	1	1	0	14	2	0	5	3	62
Suspected felony type	**122**	**87**	**59**	**2**	**7**	**19**	**15**	**5**	**2**	**0**	**0**	**0**	**1**	**0**	**0**	**4**	**1**	**7**
Other than felony type total:	**5,782**	**3,673**	**2,653**	**164**	**186**	**670**	**922**	**208**	**482**	**8**	**0**	**2**	**25**	**24**	**4**	**39**	**52**	**343**
Romantic triangle	69	49	36	3	2	8	13	0	1	1	0	0	0	0	0	2	2	3
Child killed by babysitter	30	0	0	0	0	0	0	3	19	0	0	0	0	0	0	0	1	3
Brawl due to influence of alcohol	93	48	33	5	5	5	21	3	19	0	0	0	1	0	0	0	2	7
Brawl due to influence of narcotics	59	35	26	2	1	6	3	2	4	0	0	0	0	6	0	1	1	5
Argument over money or property	133	83	62	4	6	11	25	9	11	0	0	0	6	0	0	1	1	8
Other arguments	2,889	1,747	1,270	80	108	289	616	124	216	3	0	1	11	2	2	22	22	123
Gangland killings	138	117	87	3	1	26	15	0	2	0	0	1	0	0	2	0	0	4
Juvenile gang killings	584	547	428	9	9	101	28	2	2	0	0	0	0	0	0	0	0	5
Institutional killings	15	0	0	0	0	0	0	3	7	0	0	0	0	0	0	2	0	2
Sniper attack	6	6	2	1	0	3	0	0	0	0	0	0	0	0	0	0	0	0
Other-not specified	1,766	1,041	709	57	54	221	201	62	207	4	0	1	13	16	2	14	22	183
Unknown	**4,440**	**3,313**	**2,044**	**87**	**79**	**1,103**	**379**	**136**	**138**	**2**	**0**	**0**	**26**	**8**	**0**	**25**	**31**	**382**

*The rape figures in this table are an aggregate total of the data submitted using both the revised and legacy Uniform Crime Reporting definitions.

SOURCE: "Expanded Homicide Data Table 11. Murder Circumstances by Weapon, 2013," in *Crime in the United States, 2013*, Federal Bureau of Investigation, 2014, http://www.fbi.gov/about-us/cjis/ucr/crime-in-the-u.s/2013/crime-in-the-u.s.-2013/offenses-known-to-law-enforcement/expanded-homicide/expanded_homicide_data_table_11_murder_circumstances_by_weapon_2013.xls (accessed November 20, 2014)

TABLE 5.4

Homicides, by state and type of weapon, 2013

State	Total murders[a]	Total firearms	Handguns	Rifles	Shotguns	Firearms (type unknown)	Knives or cutting instruments	Other weapons	Hands, fists, feet, etc.[b]
Alabama[c]	2	1	1	0	0	0	1	0	0
Alaska	34	12	5	3	1	3	5	13	4
Arizona	304	184	133	11	11	29	56	55	9
Arkansas	154	110	53	5	5	47	22	17	5
California	1,745	1,224	805	29	48	342	238	191	92
Colorado	174	88	46	5	3	34	34	39	13
Connecticut	86	60	34	0	0	26	12	5	9
Delaware	39	33	22	1	0	10	4	2	0
District of Columbia	103	81	30	1	1	49	12	5	5
Georgia	534	411	348	11	10	42	40	74	9
Hawaii	9	6	1	2	2	1	2	1	0
Idaho	26	15	10	1	0	4	2	5	4
Illinois[c]	433	364	352	3	2	7	41	19	9
Indiana	311	238	149	6	7	76	19	41	13
Iowa	42	18	8	1	1	8	13	6	5
Kansas	112	78	39	9	5	25	8	16	10
Kentucky	165	111	82	10	6	13	23	22	9
Louisiana	453	356	298	8	9	41	34	40	23
Maine	24	12	3	2	1	6	5	2	5
Maryland	379	268	263	0	4	1	57	29	25
Massachusetts	135	78	35	2	0	41	25	25	7
Michigan	625	440	203	16	23	198	43	106	36
Minnesota	110	60	53	0	4	3	17	24	9
Mississippi	142	110	80	3	8	19	9	14	9
Missouri	371	273	137	14	6	116	41	45	12
Montana	15	9	5	0	1	3	3	2	1
Nebraska	57	39	11	2	0	26	8	7	3
Nevada	157	87	26	0	3	58	24	36	10
New Hampshire	21	5	3	0	0	2	6	5	5
New Jersey	401	291	229	0	2	60	53	37	20
New Mexico	106	59	46	5	2	6	19	20	8
New York	648	362	290	4	19	49	136	113	37
North Carolina	452	315	204	27	8	76	50	55	32
North Dakota	11	4	0	2	0	2	4	3	0
Ohio	434	309	208	3	4	94	37	69	19
Oklahoma	191	127	108	6	8	5	24	24	16
Oregon	78	43	15	0	4	24	12	18	5
Pennsylvania	594	440	328	20	22	70	52	74	28
Rhode Island	31	18	3	0	1	14	5	6	2
South Carolina	296	224	126	9	13	76	20	33	19
South Dakota	12	3	0	0	1	2	1	4	4
Tennessee	327	223	152	11	9	51	18	67	19
Texas	1,133	760	530	34	30	166	164	129	80
Utah	49	31	25	2	1	3	6	7	5
Vermont	9	5	1	2	1	1	0	1	3
Virginia	315	225	126	3	6	90	30	40	20
Washington	155	86	47	0	7	32	29	27	13
West Virginia	54	30	17	3	3	7	2	20	2
Wisconsin	157	103	72	6	6	19	18	24	12
Wyoming	15	9	6	3	0	0	2	4	0
Guam	9	2	1	0	0	1	4	1	2
Virgin Islands	14	14	13	0	0	1	0	0	0

[a]Total number of murders for which supplemental homicide data were received.
[b]Pushed is included in hands, fists, feet, etc.
[c]Limited supplemental homicide data were received.

SOURCE: "Table 20. Murder by State, Types of Weapons, 2013," in *Crime in the United States, 2013*, Federal Bureau of Investigation, 2014, http://www.fbi.gov/about-us/cjis/ucr/crime-in-the-u.s/2013/crime-in-the-u.s.-2013/tables/table-20/table_20_murder_by_state_types_of_weapons_2013.xls (accessed November 20, 2014)

foundation that conducts research on violence in the United States, analyzes FBI data on homicides committed by family members, friends, and acquaintances and determines that "for homicides in which the victim to offender relationship could be identified, 93 percent of female victims (1,487 out of 1,594) were murdered by a male they knew." Among female murder victims who knew their killers, 61% (926) were wives or intimate partners, and 267 of these victims were "shot and killed by either their husband or intimate acquaintance during the course of an argument." More women were killed with firearms (52%) than with any other type of weapon. Of the homicides committed with firearms, almost seven out of 10 (69%) involved handguns.

TABLE 5.5

Homicide circumstances, by relationship of victim to offender, 2013[a]

Circumstances	Total murder victims	Husband	Wife	Mother	Father	Son	Daughter	Brother	Sister	Other family	Acquaintance	Friend	Boyfriend	Girlfriend	Neighbor	Employee	Employer	Stranger	Unknown
Total	12,253	108	534	128	142	230	148	99	30	245	2,660	346	137	458	127	6	2	1,281	5,572
Felony type total:	1,909	3	28	10	13	20	15	10	4	31	477	53	9	20	26	2	0	390	798
Rape[b]	20	0	1	0	0	0	0	0	0	2	6	1	0	1	1	0	0	4	4
Robbery	686	1	0	0	4	0	0	0	0	6	151	15	1	1	8	2	0	240	257
Burglary	94	0	0	0	0	0	0	0	0	1	22	1	1	2	3	0	0	23	41
Larceny-theft	16	0	2	1	0	0	0	0	0	2	5	1	0	0	2	0	0	2	7
Motor vehicle theft	27	0	0	1	2	0	0	2	2	4	2	1	1	1	1	0	0	4	6
Arson	37	0	0	0	1	0	0	4	0	2	11	0	0	0	4	0	0	6	5
Prostitution and commercialized vice	13	0	0	0	0	0	0	0	0	0	5	2	1	0	0	0	0	2	4
Other sex offenses	9	0	0	0	0	0	0	0	0	1	4	0	0	2	0	0	0	2	2
Narcotic drug laws	386	0	0	0	0	0	0	2	0	0	146	11	1	0	0	0	0	28	195
Gambling	7	0	0	0	0	0	1	0	0	0	4	0	0	0	0	0	0	0	3
Other—not specified	614	2	25	8	6	20	14	2	2	13	121	21	5	13	7	0	0	79	274
Suspected felony type	122	1	5	4	2	2	2	0	1	2	16	2	0	6	1	0	0	6	72
Other than felony type total:	5,782	89	410	72	102	173	94	72	19	163	1,644	210	109	348	83	4	2	617	1,571
Romantic triangle	69	0	2	0	1	0	0	0	0	0	36	8	3	9	0	0	0	3	7
Child killed by babysitter	30	0	0	0	0	2	0	0	0	2	25	0	0	0	1	0	0	0	0
Brawl due to influence of alcohol	93	0	4	1	2	0	1	1	0	6	27	14	0	3	1	0	0	21	11
Brawl due to influence of narcotics	59	0	1	0	1	0	0	0	0	1	28	5	0	2	1	0	0	6	13
Argument over money or property	133	0	4	3	0	1	0	3	1	6	59	14	1	6	4	0	0	11	20
Other arguments	2,889	65	298	46	74	39	14	56	10	98	891	117	90	273	56	2	2	250	508
Gangland killings	138	0	1	0	0	0	0	0	0	0	40	0	0	0	0	0	0	24	73
Juvenile gang killings	584	0	0	0	0	0	0	0	0	0	106	1	0	0	1	0	0	80	396
Institutional killings	15	0	0	0	0	0	0	0	0	0	10	0	0	0	0	0	0	3	2
Sniper attack	6	0	0	0	0	0	0	0	0	0	2	0	0	0	0	0	0	2	2
Other—not specified	1,766	24	100	22	24	130	79	12	8	50	420	51	14	55	19	2	0	217	539
Unknown	4,440	15	91	42	25	35	37	17	6	49	523	81	19	84	17	0	0	268	3,131

[a]Relationship is that of victim to offender.

[b]The rape figures in this table are an aggregate total of the data submitted using both the revised and legacy Uniform Crime Reporting definitions.

Note: The relationship categories of husband and wife include both common-law and ex-spouses. The categories of mother, father, sister, brother, son, and daughter include stepparents, stepchildren, and stepsiblings. The category of acquaintance includes homosexual relationships and the composite category of other known to victim.

SOURCE: "Expanded Homicide Data Table 10. Murder Circumstances by Relationship, 2013," in *Crime in the United States, 2013*, Federal Bureau of Investigation, 2014, http://www.fbi.gov/about-us/cjis/ucr/crime-in-the-u.s/2013/crime-in-the-u.s.-2013/offenses-known-to-law-enforcement/expanded-homicide/expanded_homicide_data_table_10_murder_circumstances_by_relationship_2013.xls (accessed November 20, 2014)

The Criminal Advantages of Guns

A gun offers a criminal several advantages over other weapons. A gun offender can keep a greater physical distance from the victim to ensure his or her own safety and increase the chances of escaping. Because guns can kill from a great distance, they are also the most effective weapon against well-guarded targets. Perhaps the best example of this phenomenon is the use of firearms in political assassinations or assassination attempts. Also, it is unlikely that a felon would try to rob a well-guarded institution with many customers and multiple exits, such as a bank, while wielding a club or knife. A gun allows the offender to maintain not just physical distance but also psychological distance, keeping the confrontation more impersonal and minimizing the emotional involvement, a factor that some people claim leads to more killings than would otherwise occur. Control over potential victims can be easier to maintain with a gun; victims are less likely to run from a gun-carrying offender than from an offender who brandishes other types of weapons, such as knives, for fear of being shot from a distance.

Finally, multiple-victim homicides are particularly dependent on access to firearms. The higher the number of homicide victims, the greater the likelihood that a firearm was the weapon used. In 2011, 67.1% of all homicides involved a firearm, and the percentage of single-victim homicides involving a firearm was 66.5%. (See Table 5.6.) By comparison, 77.3% of two-victim homicides, 82.3% of three-victim homicides, and 90.8% of homicides in which there were four or more victims involved firearms. As these statistics show, mass murders are almost always mass shootings. Firearms, unlike other commonly available weapons, make it possible to kill a large number of people in a matter of seconds or minutes. Although perpetrators are usually apprehended or commit suicide, there is an inevitable lag time after the perpetrators begin shooting but before they are met with deadly force.

Mass Shootings in the United States

Table 5.7 provides an overview of notable multiple or mass shootings that have occurred in the United States since 1949. Many of the deadliest mass shootings in U.S. history have occurred at schools, most often high schools and college campuses. These include the 1999 Columbine shootings in which Dylan Klebold (1982–1999) and Eric Harris (1981–1999) killed 13 people and wounded 23 others at Columbine High School in Littleton, Colorado, before shooting and killing themselves; and the 2007 massacre at Virginia Polytechnic Institute and State University, in which Seung-Hui Cho (1984–2007) used two handguns to kill 32 students and faculty members and injure 15 others before fatally shooting himself. Mass shootings in U.S. schools had become grimly common in the United States by 2012, but that year brought a new form of mass-shooting terror, when 20-year-old Adam Lanza (1992?–2012) targeted an elementary school, killing 20 first graders at Sandy Hook Elementary School in Newtown, Connecticut, along with his mother and six school staff members before committing suicide. Other notable mass shootings have occurred in the workplace, including the 1986 shooting at the Edmond, Oklahoma, post office, in which Patrick Sherrill (1941–1986) killed 14 coworkers and injured six and then himself; and the Fort Hood, Texas, shooting of 2009, in which Major Nidal Malik Hasan killed 12 fellow servicemembers and wounded 31.

Table 5.7 is not a comprehensive list of mass shootings in the United States, but it does reflect the deadliest attacks as well as their approximate relative distribution over time. The list dates to 1949, when one of the first high-profile multiple shootings in the United States occurred, but incidents of this type were uncommon prior to the 1990s. Since that time, the rate at which mass

TABLE 5.6

Homicides, by number of victims and weapon type, 2011

| Number of homicide victims | Number of homicide incidents | Total | Percent of homicides involving a— | | | |
| | | | Firearm | | | |
			Any firearm	Handgun	Other firearm*	Other weapon
Total	13,750	100%	67.1	49.4	17.7	32.9
1 victim	13,050	100%	66.5	49.3	17.2	33.5
2 victims	565	100%	77.3	52.0	25.3	22.7
3 victims	110	100%	82.3	47.1	35.2	17.7
4 or more victims	25	100%	90.8	44.2	46.6	9.2

*Includes rifles, shotguns, and firearms of unspecified type, including automatic weapons.
Note: Due to limitations of the data, the incident count presented above may not accurately reflect the total number of unique homicide incidents in the United States.

SOURCE: Erica L. Smith and Alexia Cooper, "Table 5. Homicides in the U.S., by the Number of Victims Killed and Weapon Type, 2011," in *Homicide in the U.S. Known to Law Enforcement, 2011*, U.S. Department of Justice, Bureau of Justice Statistics, December 2013, http://www.bjs.gov/content/pub/pdf/hus11 .pdf (accessed July 15, 2014)

TABLE 5.7

Notable multiple shootings, 1949–2014

2014

Elliot Rodger, 22, stabbed three men in his apartment before embarking on a rampage in the streets of Isla Vista, California, near the campus of University of California, Santa Barbara. He ultimately killed six and injured 13, using three handguns as well as his automobile, before committing suicide.

2014

Enlisted soldier Ivan Lopez, 34, killed three people and injured 16 at Fort Hood, TX, the site of a previous mass shooting in 2009 (see below), before killing himself.

2013

Aaron Alexis, 34, a former member of the U.S. Navy who had been discharged from service in 2011, killed 12 people and wounded three at the Washington, D.C., Navy Yard complex, using three weapons including an AR-15 assault rifle, before being killed by police and navy security officials.

2012

Adam Lanza, 20, killed 27 people, beginning with his mother, before he moved on to Sandy Hook Elementary School in Newtown, Connecticut, where he killed 20 first-graders and six staff members, before taking his own life. Lanza, a firearms enthusiast along with his mother, used an assault-style weapon as well as pistols in the attacks, weapons that he had obtained at home.

2012

Wade Michael Page, 40, killed six people and wounded four at a Sikh temple in Oak Creek, WI. Page, who took his own life at the scene after being wounded by police gunfire, used a 9mm semiautomatic handgun in the spree.

2012

Shortly after the beginning of a midnight screening of the film *The Dark Knight Rises*, a gunman in protective gear ignited tear gas canisters in a theater auditorium in Aurora, CO, and then opened fire on the disoriented audience members, killing 12 and wounding 58 others. The shooter, James Holmes, 24, was arrested outside the theater moments later with a .223-caliber assault rifle, a 12-gauge shotgun, and a .40-caliber handgun.

2012

One L. Goh, 43, killed seven and wounded three at Oikos University, a Korean Christian college in Oakland, California. Goh, a former student at the college, used a legally purchased handgun to commit the crimes, and he turned himself into authorities. He was later found mentally unfit for trial and was admitted to a psychiatric hospital for treatment.

2011

U.S. Representative Gabrielle Giffords (D-AZ) was among the injured victims when Jared Loughner, 22, opened fire with a 9mm semiautomatic handgun at a political event near Tucson, AZ. Six people were killed and 13 were wounded in the incident, with an additional victim sustaining injuries in the ensuing chaos.

2010

Christopher Bryan Speight, a 39-year-old security guard, shot and killed his sister and brother-in-law, two other adults, three teenagers, and a four-year old in an Appomattox, VA, house and surrounding yard. Speight used a rifle in the attack.

2009

At the American Civic Association in Binghampton, NY, a services center for immigrants, Jiverly Antares Wong, 41, entered the building and opened fire without saying a word, killing 13 and wounding four others. Many of the victims were immigrants participating in an English class at the center. Wong used two handguns in the attack.

2009

In a spree that lasted less than an hour, 28-year old gunman Michael Kenneth McLendon killed 11 people and four dogs in two communities in Geneva County, AL, before killing himself. Among those fatally shot were his mother, grandmother, and several other family members; five additional victims survived their wounds.

2009

Major Nidal Malik Hasan, a 39-year-old practicing psychiatrist at the Darnall Army Medical Center at Fort Hood, TX, opened fire at the base's processing center. Using two handguns, Hasan killed 12 and wounded 31.

2008

On the campus of Northern Illinois University in DeKalb, Illinois, former graduate student Steven Phillip Kazmierczak, 27, entered an auditorium lecture hall and began shooting at the teacher and students assembled for the class. Armed with several weapons, including a 12-gauge shotgun and a 9mm Glock semiautomatic pistol, he killed five students and wounded 21 others before killing himself.

2008

At the Kirkwood, MO, city hall, Charles Lee "Cookie" Thornton, 52, stormed a city council meeting, shooting and killing the public works director, two city council members, and two police officers. He seriously injured the city's mayor and wounded a newspaper reporter. Police at the scene shot and killed Thornton, who was said to have a long-standing feud with the city.

2007

In the deadliest shooting rampage by a single gunman in U.S. history, Seung-Hui Cho, a 23-year-old college student, killed 32 students and faculty members at Virginia Polytechnic Institute and State University. In the shooting spree, 15 people were also wounded. Cho then shot and killed himself.

2005

Jeffrey Weise, 16, killed his grandfather and his grandfather's girlfriend at their home on the Red Lake Indian Reservation in Red Lake, Minnesota, before driving to Red Lake Senior High School, where he had formerly been a student, killing seven and wounding five before committing suicide. Weise used a shotgun and two handguns in the attacks.

2003

Wielding a semiautomatic pistol, Jonathan Russell, 25, killed three of his co-workers and injured five others before killing himself outside of Modine Manufacturing Co., in Jefferson City, MO.

2002

John Allen Muhammad, 41, who qualified as an expert marksman with the M-16 in the U.S. Army, and his stepson, 17-year-old John Lee Malvo, killed 10 people and wounded three others in Washington, DC, and vicinity, in a three-week killing spree with a .232-caliber Bushmaster XM15 semiautomatic rifle.

2000

In Queens, NY, two young gunmen bound, gagged, and shot seven employees in a Wendy's restaurant with a .380-caliber gun. Five of the workers were killed.

shootings occur has accelerated. According to Mark Follman, Gavin Aronsen, and Deanna Pan, in "A Guide to Mass Shootings in America" (MotherJones.com, May 24, 2014), since 1982 there have been at least 69 mass shootings in 30 states, 32 of which occurred between 2006 and 2014.

TABLE 5.7

Notable multiple shootings, 1997–2014 [CONTINUED]

2000

Richard Glassel, 61, armed with three handguns, an AR-15 assault rifle, and hundreds of rounds of ammunition, shot and killed two women and injured three other people during a homeowners association meeting in Arizona.

1999

In Los Angeles, five were wounded and a postal worker was fatally shot by Buford O. Furrow Jr., 37, at the North Valley Jewish Community Center.

1999

In Atlanta, GA, Mark Barton, 44, killed nine people and wounded 13 at two brokerage firms before killing himself.

1999

Benjamin Nathaniel Smith, 21, killed two people and injured nine in a three-day rampage through Indiana and Illinois, before shooting himself.

1999

At Columbine High School in Littleton, CO, Dylan Klebold, 17, and Eric Harris, 18, killed 12 fellow students and a teacher and wounded 23 others, before shooting and killing themselves. Klebold and Harris each used a shotgun and a handgun for the killings; they also planted numerous explosive devices at the school, but the explosives failed to detonate.

1998

In Springfield, OR, 15-year-old Kip Kinkel fired more than 50 rounds from a .22-caliber semiautomatic rifle into a high school cafeteria. Two male students died and 23 other students were injured. The boy also shot and killed his parents.

1998

Four middle school students and a teacher were killed and 10 other students were injured in Jonesboro, AR, when a 13-year-old and an 11-year-old shot at the school from a nearby wooded area.

1997

Using a gun, Ali Hassan Abu Kamal, 69, killed one person and injured six others before taking his own life on the observation deck of the Empire State Building in Manhattan.

1991

George Hennard, 35, crashed his vehicle into a Luby's Cafeteria in Killeen, TX, exited his truck, and opened fire on the restaurant's customers and staff with two handguns. He killed 23 people and wounded 20 more before committing suicide.

1986

Patrick Sherrill, 44, killed 14 of his coworkers and injured another six at the Edmond, OK, post office where he worked, before killing himself. The attack, in which Sherrill used three handguns, is the deadliest act of workplace violence in U.S. history.

1984

James Huberty, 41, killed 21 (including five children) and injured 19 at a McDonald's restaurant in San Diego, CA, using two handguns and a shotgun, before being killed by a police sniper.

1966

Charles Joseph Whitman, 25, killed his wife and mother in Austin, TX, before moving on to the campus of the University of Texas, where he killed three people inside a campus tower and then murdered 11 more people with a rifle from his vantage point on the tower's 28th-floor observation deck. In all, Whitman killed 16 and injured 32 before he was killed by a police officer.

1949

Howard Unruh, 28, killed 13 people (three of whom were children) and injured three while walking through his Camden, NJ, neighborhood and firing on his neighbors with a handgun. Unruh was found criminally insane and lived in a New Jersey psychiatric hospital for the remainder of his life, dying at age 88 in 2009.

SOURCE: Created by Sandra Alters, Laurie DiMauro, and Mark Lane for Gale, © 2014.

POLICE DEATHS AND INJURIES

The National Law Enforcement Officers Memorial Fund notes in "Causes of Law Enforcement Deaths" (April 14, 2014, http://www.nleomf.org/facts/officer-fatalities-data/causes.html) that 548 federal, state, and local law enforcement officers had been killed in firearms-related incidents between 2004 and 2013. Shootings were the most common cause of death for law enforcement officers, accounting for 37% of the total 1,501 law enforcement deaths over the course of that decade. Auto crashes, accounting for 434 deaths (29%), were the second-leading cause of death for law enforcement officers.

In *Law Enforcement Officers Killed and Assaulted, 2012* (2013, http://www.fbi.gov/about-us/cjis/ucr/leoka/2012), the FBI indicates that 48 law enforcement officers were killed in the line of duty in 2012, 44 of them with firearms. (See Table 5.8.) Handguns were used in 32 of the killings, rifles in seven, and shotguns in three. The

South was the U.S. region in which law enforcement officers were the most likely to be killed with a firearm while on duty, accounting for 18 of the 44 total deaths in 2012, more than double the number of deaths in the West, the region with the second-highest total.

Table 5.9 details the circumstances under which federal, state, and local police officers were killed in the line of duty between 2003 and 2012. Of the 535 officers killed over the course of the decade, 124 (23.2%) were killed in arrest situations and 115 (21.5%) were killed in ambush situations. Another 96 (17.9%) were killed during traffic stops, 64 (12%) while responding to disturbance calls, 60 (11.2%) while investigating suspicious people or circumstances, and 43 (8%) in high-risk tactical situations.

Similar numbers of law enforcement officers are also killed accidentally in the line of duty. As Table 5.10 shows, 667 officers lost their lives on duty as a result of

TABLE 5.8

Law enforcement officers killed, by type of weapon and region, 2012

Region	Total	Total firearms	Handgun	Rifle	Shotgun	Type of firearm not reported	Knife or other cutting instrument	Bomb	Blunt instrument	Personal weapons	Vehicle	Other
Number of victim officers	**48**	**44**	**32**	**7**	**3**	**2**	**1**	**0**	**0**	**1**	**2**	**0**
Northeast	6	6	4	1	0	1	0	0	0	0	0	0
Midwest	6	6	4	0	2	0	0	0	0	0	0	0
South	22	18	14	3	1	0	1	0	0	1	2	0
West	8	8	5	3	0	0	0	0	0	0	0	0
Puerto Rico and other outlying areas	6	6	5	0	0	1	0	0	0	0	0	0

SOURCE: "Table 29. Law Enforcement Officers Feloniously Killed, Region by Type of Weapon, 2012," in *Law Enforcement Officers Killed and Assaulted, 2012*, Federal Bureau of Investigation, 2013, http://www.fbi.gov/about-us/cjis/ucr/leoka/2012/tables/table_29_leos_fk_region_by_type_of_weapon_2012.xls (accessed July 15, 2014)

TABLE 5.9

Law enforcement officers killed, by circumstance at scene of incident, 2003–12

Circumstance	Total	2003	2004	2005	2006	2007	2008	2009	2010	2011	2012
Number of victim officers											
Total	535	52	57	55	48	58	41	48	56	72	48
Disturbance call											
Total	64	10	10	7	8	5	1	6	6	7	4
Disturbance (bar fight, person with firearm, etc.)	31	5	1	2	6	3	1	4	2	5	2
Domestic disturbance (family quarrel, etc.)	33	5	9	5	2	2	0	2	4	2	2
Arrest situation											
Total	124	8	13	8	12	17	9	8	14	23	12
Burglary in progress/pursuing burglary suspect	12	1	2	1	0	1	2	1	3	0	1
Robbery in progress/pursuing robbery suspect	44	1	7	4	6	7	1	3	6	5	4
Drug-related matter	8	1	0	0	2	1	1	0	1	0	2
Attempting other arrest	60	5	4	3	4	8	5	4	4	18	5
Civil disorder (mass disobedience, riot, etc.)											
Total	0	0	0	0	0	0	0	0	0	0	0
Handling, transporting, custody of prisoner											
Total	14	2	1	1	1	1	1	2	1	1	3
Investigating suspicious person/circumstance											
Total	60	4	7	7	6	4	7	4	8	5	8
Ambush situation											
Total	115	9	15	8	10	16	6	15	15	15	6
Entrapment/premeditation	41	6	6	4	1	9	1	6	2	2	4
Unprovoked attack	74	3	9	4	9	7	5	9	13	13	2
Investigative activity (surveillance, search, interview, etc.)											
Total	13	2	0	4	0	1	2	0	2	1	1
Handling person with mental illness											
Total	6	0	2	2	1	0	0	0	0	0	1
Traffic pursuit/stop											
Total	96	14	6	15	8	11	8	8	7	11	8
Felony vehicle stop	34	4	0	5	0	5	5	2	3	6	4
Traffic violation stop	62	10	6	10	8	6	3	6	4	5	4
Tactical situation (barricaded offender, hostage taking, high-risk entry, etc.)											
Total	43	3	3	3	2	3	7	5	3	9	5

SOURCE: "Table 19. Law Enforcement Officers Feloniously Killed, Circumstance at Scene of Incident, 2003–2012," in *Law Enforcement Officers Killed and Assaulted, 2012*, Federal Bureau of Investigation, 2013, http://www.fbi.gov/about-us/cjis/ucr/leoka/2012/tables/table_19_leos_fk_circumstance_at_scene_of_incident_2003-2012.xls (accessed July 15, 2014)

TABLE 5.10

Law enforcement officers accidentally killed, by circumstance at scene of incident, 2003–12

Circumstance	Total	2003	2004	2005	2006	2007	2008	2009	2010	2011	2012
Number of victim officers											
Total	667	81	82	67	66	83	68	48	72	53	47
Automobile accident											
Total	394	50	48	39	38	49	39	34	45	30	22
Motorcycle accident											
Total	64	10	10	4	8	6	6	3	7	4	6
Aircraft accident											
Total	21	1	3	2	3	3	2	1	2	1	3
Struck by vehicle											
Total	102	10	10	11	13	12	13	7	11	5	10
Traffic stop, roadblock, etc.	38	6	3	5	4	7	1	3	4	3	2
Directing traffic, assisting motorist, etc.	64	4	7	6	9	5	12	4	7	2	8
Accidental shooting											
Total	31	2	4	4	4	4	2	2	3	4	2
Crossfire, mistaken for subject, firearm mishap	22	1	2	2	3	4	2	2	1	3	2
Training session	3	0	1	1	0	0	0	0	1	0	0
Self-inflicted, cleaning mishap (not apparent or confirmed suicide)	6	1	1	1	1	0	0	0	1	1	0
Drowning											
Total	15	4	3	2	0	2	1	0	0	3	0
Fall											
Total	13	2	1	3	0	1	0	0	1	2	3
Other accidental											
Total	27	2	3	2	0	6	5	1	3	4	1

SOURCE: "Table 61. Law Enforcement Officers Accidentally Killed, Circumstance at Scene of Incident, 2003–2012," in *Law Enforcement Officers Killed and Assaulted, 2012*, Federal Bureau of Investigation, 2013, http://www.fbi.gov/about-us/cjis/ucr/leoka/2012/tables/table_61_leos_ak_circumstance_at_scene_of_incident_2003-2012.xls (accessed July 15, 2014)

accidents between 2003 and 2012, a majority of them (394, or 59.1%) in automobile accidents. Accidental shootings accounted for relatively few officer deaths (31, or 4.6%) over the course of the decade.

Officers Assaulted

Table 5.11 shows the number of assaults on federal officers between 2008 and 2012. During this period there were 8,587 assaults on federal law enforcement officers. Firearms were not the most common type of weapon used in assaults; they were used in 442 incidents, which accounted for only 5.1% of the total. By comparison, personal weapons (e.g., fists) were used in 2,647 (30.8%) cases. Of the 104 firearm assaults on federal officers reported in 2012, one resulted in death, seven resulted in injuries, and 71 in no injuries.

In 2012, 52,901 state and local law enforcement officers were assaulted. (See Table 5.12.) Again, firearms were the type of weapons used in only a small number of cases, accounting for 2,259 (4.3%) of the total. Personal weapons such as fists and feet accounted for the overwhelming majority of these assaults (42,408, or 80.2%).

NONFATAL VIOLENT CRIMES INVOLVING FIREARMS

In *Firearm Violence, 1993–2011* (May 2013, http://www.bjs.gov/content/pub/pdf/fv9311.pdf), Michael Planty and Jennifer L. Truman of the BJS indicate that the incidence of nonfatal firearm violence, like that of the firearm homicide rate and the homicide rate in general, peaked during the mid-1990s and then fall markedly thereafter. Nonfatal firearm violence peaked in 1994 at approximately 1.3 million incidents, falling to a low of 331,600 incidents in 2008 before rising slightly in the following years. (See Figure 5.7.)

The firearm violence rate (the incidence of firearm violence per each 1,000 people aged 12 years and older) has been steady throughout the 21st century: it was 2 per 1,000 in 2003, 1.8 per 1,000 in 2011, and 1.8 per 1,000 in 2012. (See Table 5.13.) Guns were used in 6.6% of all violent crime incidents in 2012, up slightly from 2003 (5.6%) and down slightly from 2011 (7.7%). Two-thirds (66.4%) of firearm crime incidents were reported to police in 2012, down from 73.5% in both 2003 and 2011.

TABLE 5.11

Assaults on federal officers, by extent of injury and type of weapon, 2008–12

Year	Extent of injury	Total	Firearm	Knife or other cutting instrument	Bomb	Blunt instrument	Personal weapons	Vehicle	Other
Number of victim officers	Total	8,587	442	95	31	105	2,647	476	4,791
2008	Total	1,347	71	11	0	11	400	85	769
	Killed	2	1	0	0	0	0	1	0
	Injured	188	2	2	0	6	125	19	34
	Not injured	1,157	68	9	0	5	275	65	735
2009	Total	1,807	89	16	0	21	464	109	1,108
	Killed	1	1	0	0	0	0	0	0
	Injured	181	6	3	0	8	83	18	63
	Not injured	1,625	82	13	0	13	381	91	1,045
2010[a]	Total	1,886	86	17	0	16	387	76	1,304
	Killed	1	1	0	0	0	0	0	0
	Injured	351	12	0	0	3	117	18	201
	Not injured	1,534	73	17	0	13	270	58	1,103
2011	Total	1,689	92	18	15	19	663	102	780
	Killed	3	3	0	0	0	0	0	0
	Injured	254	7	4	0	6	147	20	70
	Not injured	1,432	82	14	15	13	516	82	710
2012	Total	1,858	104	33	16	38	733	104	830
	Killed	1	1	0	0	0	0	0	0
	Injured	206	7	3	0	5	125	7	59
	Not injured	1,096	71	30	1	22	414	47	511
	Extent of injury not reported[b]	555	25	0	15	11	194	50	260

[a]Prior to 2010, data were not collected from the U.S. Customs and Border Protection, Office of Air and Marine.
[b]For 2012, extent of injury data were not reported by the U.S. Customs and Border Protection, Office of Border Patrol.

SOURCE: "Table 77. Federal Law Enforcement Officers Killed and Assaulted, Extent of Injury of Victim Officer by Type of Weapon, 2008–2012," in *Law Enforcement Officers Killed and Assaulted, 2012*, Federal Bureau of Investigation, 2013, http://www.fbi.gov/about-us/cjis/ucr/leoka/2012/tables/table_77_fed_leos_k_and_a_extent_of_injury_of_victim_officer_by_type_of_weapon_2008-2012.xls (accessed July 15, 2014)

TABLE 5.12

Law enforcement officers assaulted, by circumstance at scene of incident and type of weapon, 2012

Circumstance	Total	Percent distribution	Firearm Total	Firearm Percent distribution	Knife or other cutting instrument Total	Knife or other cutting instrument Percent distribution	Other dangerous weapon Total	Other dangerous weapon Percent distribution	Personal weapons Total	Personal weapons Percent distribution
Number of victim officers	52,901	100.0	2,259	4.3	893	1.7	7,341	13.9	42,408	80.2
Disturbance call	17,205	100.0	753	4.4	380	2.2	1,692	9.8	14,380	83.6
Burglary in progress/pursuing burglary suspect	760	100.0	42	5.5	11	1.4	139	18.3	568	74.7
Robbery in progress/pursuing robbery suspect	463	100.0	100	21.6	13	2.8	80	17.3	270	58.3
Attempting other arrest	8,057	100.0	207	2.6	109	1.4	843	10.5	6,898	85.6
Civil disorder (mass disobedience, riot, etc.)	722	100.0	7	1.0	3	0.4	193	26.7	519	71.9
Handling, transporting, custody of prisoner	7,173	100.0	64	0.9	35	0.5	605	8.4	6,469	90.2
Investigating suspicious person/circumstance	4,915	100.0	307	6.2	101	2.1	674	13.7	3,833	78.0
Ambush situation	267	100.0	104	39.0	12	4.5	48	18.0	103	38.6
Handling person with mental illness	1,353	100.0	80	5.9	82	6.1	163	12.0	1,028	76.0
Traffic pursuit/stop	4,450	100.0	280	6.3	34	0.8	1,548	34.8	2,588	58.2
All other	7,536	100.0	315	4.2	113	1.5	1,356	18.0	5,752	76.3

Note: Because of rounding, the percentages may not add to 100.0.

SOURCE: "Table 73. Law Enforcement Officers Assaulted, Circumstance at Scene of Incident by Type of Weapon and Percent Distribution, 2012," in *Law Enforcement Officers Killed and Assaulted, 2012*, Federal Bureau of Investigation, 2013, http://www.fbi.gov/about-us/cjis/ucr/leoka/2012/tables/table_73_leos_asltd_circum_at_scene_of_incident_by_type_of_weapon_and_percent_distribution_2012.xls (accessed July 15, 2014)

Table 5.14 shows levels of firearm violence between 1994 and 2011 by the type of firearm used. Although the number of violent incidents involving firearms has fallen drastically, handguns have consistently accounted for the overwhelming majority of firearm violence during this period. Handguns accounted for between 84.2% (in 1997)

FIGURE 5.7

Nonfatal firearm violence, 1993–2011

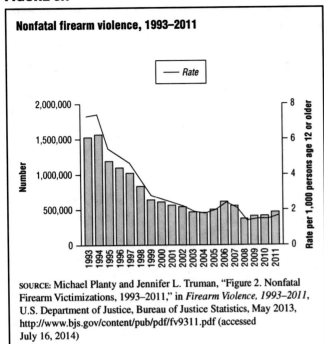

SOURCE: Michael Planty and Jennifer L. Truman, "Figure 2. Nonfatal Firearm Victimizations, 1993–2011," in *Firearm Violence, 1993–2011*, U.S. Department of Justice, Bureau of Justice Statistics, May 2013, http://www.bjs.gov/content/pub/pdf/fv9311.pdf (accessed July 16, 2014)

TABLE 5.13

Firearm violence, 2003, 2011, and 2012

	2003	2011	2012
Firearm incidents	385,040	415,160	427,700
Firearm victimizations	467,350	467,930	460,720
Rate of firearm violence*	2.0	1.8	1.8
Reported to the police*	1.4	1.3	1.2
Not reported to the police*	0.5	0.5	0.6
Percent of all violent incidents	5.6%	7.7%	6.6%
Percent reported to the police	73.5%	73.5%	66.4%

*Per 1,000 persons age 12 or older.
Note: Includes violent incidents and victimizations in which the offender had, showed, or used a firearm.

SOURCE: Jennifer Truman, Lynn Langton, and Michael Planty, "Table 2. Firearm Violence, 2003, 2011, and 2012," in *Criminal Victimization, 2012*, U.S. Department of Justice, Bureau of Justice Statistics, October 2013, http://www.bjs.gov/content/pub/pdf/cv12.pdf (accessed July 16, 2014)

and 92.6% (2010) of all nonfatal crimes involving firearms, and no clear upward or downward trend in this percentage is evident. Notably, the raw number of nonfatal firearm crimes for both handguns and other types of firearms has fallen dramatically since 1994, whereas among homicides committed with other types of firearms it has remained relatively constant.

Characteristics of Nonfatal Firearm Violence Victims

Nonfatal firearm violence rates differ by sex, but the rates for males and females converged between 1994 and 2011 as violence declined dramatically among males. (See Figure 5.8.) The rate of nonfatal firearm violence for males declined from 10.1 per 1,000 to 1.9 per 1,000 during this time, for a decline of 81%. Over the same period the rate of nonfatal firearm violence for females declined from 4.7 per 1,000 to 1.6 per 1,000, for a decline of 67%. Most of the decline for both males and females came between 1994 and 1998.

Rates of nonfatal firearm violence differed dramatically by race and Hispanic origin in 1994 but the rates for different demographic groups converged over the course of the following 17 years. (See Figure 5.9.) The rates of nonfatal firearm violence for non-Hispanic African Americans declined from over 16 per 1,000 in 1994 to 2.8 per 1,000 in 2011, while the rates for Hispanics declined from almost 13 per 1,000 to 2.2 per 1,000. Meanwhile, the rates for whites fell from just over 5 per 1,000 to 1.4 per 1,000. Thus, although the rates for non-Hispanic African Americans and Hispanics were consistently higher than those for whites, all three groups saw their nonfatal firearm violence rates fall at comparable levels: 83% for both non-Hispanic African Americans and Hispanics and 74% for whites.

Nonfatal firearm violence rates also converged for urban, rural, and suburban locations between 1994 and 2011. (See Figure 5.10.) The rates for all three types of locations declined between 76% and 78% during this period. Urban rates of nonfatal firearm violence, at 2.5 per 1,000, remained slightly higher in 2011 than suburban (1.4) and rural (1.2) rates. As Table 5.15 shows, however, nonfatal firearm violence rates differed significantly depending on the size of the city. Cities of between 100,000 and 249,999 saw rates decline more dramatically than larger cities, from 7 per 1,000 in 1997 to 1.3 in 2011. Thus, the rates in these smaller cities were indistinguishable from the rates in suburban and rural areas. Rates fell markedly in cities of 250,000 to 499,999 (from 10.3 to 3.9), in cities of 500,000 to 999,999 (from 7.3 to 4.6), and in cities of 1 million or more (from 7.3 to 3.2), but in these midsized or larger cities, the rates were well above the rates in suburban and rural areas.

The likelihood that one will be a victim of nonfatal firearm violence is greatest between the ages of 18 and 24 years. As Table 5.16 shows, this has been true for all years between 1993 and 2011, except 2008. Even so, the rates for all age groups fell during this period. The nonfatal firearm violence rate for children aged 12 to 17 years fell from 11.4 per 1,000 in 1994 to 1.4 per 1,000 in 2011; for those aged 18 to 24 years, from 18.1 to 5.2; for those aged 25 to 34 years, from 8.7 to 2.2; for those aged 35 to 49 years, from 6.3 to 1.4; and for those aged 50 years and older, from 1.6 to 0.7.

Of the 2.2 million victims of nonfatal firearm violence between 2007 and 2011, 77% escaped the incidents in which they were involved without physical injury. (See Table 5.17.) Twenty-three percent received some sort of

TABLE 5.14

Firearm violence, by type of firearm, 1994–2011

| | Homicide | | | | Nonfatal violence | | | | | |
| | Handgun | | Other firearm* | | Handgun | | Other firearm* | | Gun type unknown | |
Year	Annual number	Percent	Annual number	Percent	Average annual number	Percent	Average annual number	Percent	Average annual number	Percent
1994	13,510	82.7%	2,830	17.3%	1,387,100	89.5%	150,200	9.7%	11,700!	0.8%!
1995	12,090	81.9	2,670	18.1	1,240,200	89.8	132,800	9.6	7,700!	0.6!
1996	10,800	81.1	2,510	18.9	999,600	87.1	141,000	12.3	6,400!	0.6!
1997	9,750	78.8	2,630	21.2	894,200	84.2	159,800	15.0	8,400!	0.8!
1998	8,870	80.4	2,160	19.6	783,400	84.3	141,100	15.2	5,300!	0.6!
1999	8,010	78.8	2,150	21.2	659,600	89.4	74,100	10.0	4,500!	0.6!
2000	8,020	78.6	2,190	21.4	555,800	88.8	65,300	10.4	4,500!	0.7!
2001	7,820	77.9	2,220	22.1	506,600	86.3	65,900	11.2	14,100!	2.4!
2002	8,230	75.8	2,620	24.2	471,600	85.5	63,200	11.5	16,700!	3.0!
2003	8,890	80.3	2,180	19.7	436,100	86.6	53,200	10.6	14,400!	2.9!
2004	8,330	78.0	2,350	22.0	391,700	84.8	53,400	11.6	16,900!	3.7!
2005	8,550	75.1	2,840	24.9	410,600	85.5	56,200	11.7	13,200!	2.8!
2006	9,060	77.0	2,700	23.0	497,400	89.0	47,600	8.5	14,000!	2.5!
2007	8,570	73.6	3,080	26.4	509,700	87.2	65,600	11.2	9,300!	1.6!
2008	7,930	71.8	3,120	28.2	400,700	86.5	57,400	12.4	5,000!	1.1!
2009	7,370	71.3	2,970	28.7	348,700	89.2	37,600	9.6	4,400!	1.1!
2010	6,920	69.6	3,030	30.4	382,100	92.6	26,700	6.5	3,800!	0.9!
2011	7,230	72.9	2,690	27.1	389,400	88.3	49,700	11.3	2,100!	0.5!

*Includes rifle, shotgun, and other types of firearms.
!Interpret with caution. Estimate based on 10 or fewer sample cases, or coefficient of variation is greater than 50%.
Note: Nonfatal violence data based on 2-year rolling averages beginning in 1993. Homicide data are presented as annual estimates.

SOURCE: Michael Planty and Jennifer L. Truman, "Table 3. Criminal Firearm Violence, by Type of Firearm, 1994–2011," in *Firearm Violence, 1993–2011*, U.S. Department of Justice, Bureau of Justice Statistics, May 2013, http://www.bjs.gov/content/pub/pdf/fv9311.pdf (accessed July 16,2014)

FIGURE 5.8

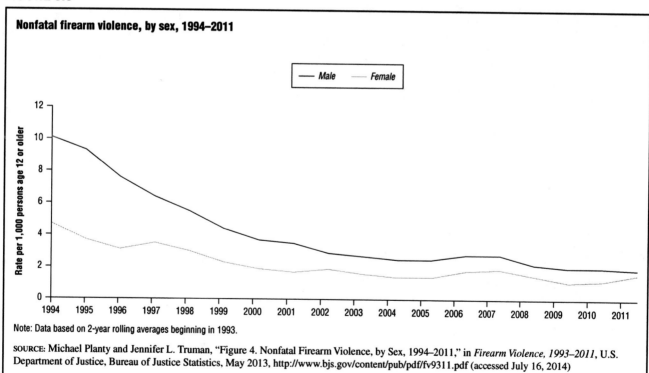

Nonfatal firearm violence, by sex, 1994–2011

Note: Data based on 2-year rolling averages beginning in 1993.

SOURCE: Michael Planty and Jennifer L. Truman, "Figure 4. Nonfatal Firearm Violence, by Sex, 1994–2011," in *Firearm Violence, 1993–2011*, U.S. Department of Justice, Bureau of Justice Statistics, May 2013, http://www.bjs.gov/content/pub/pdf/fv9311.pdf (accessed July 16, 2014)

injury, including 6.7% whose injuries were classified as serious, 16.1% whose injuries were classified as minor, and 0.2% who suffered rape without being injured by the firearms themselves. Among the victims who were injured, 72.5% received some form of treatment, including 81.6% who received treatment in medical facilities.

FIGURE 5.9

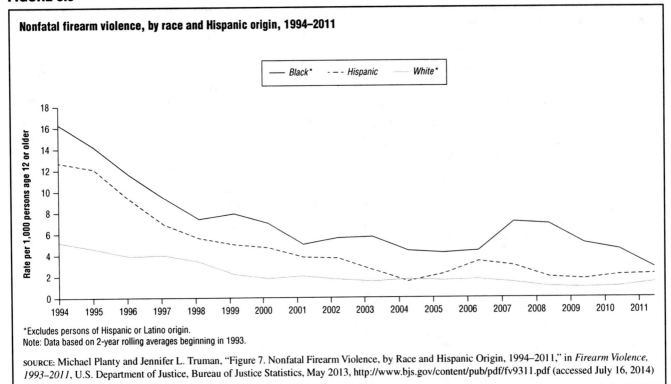

Nonfatal firearm violence, by race and Hispanic origin, 1994–2011

*Excludes persons of Hispanic or Latino origin.
Note: Data based on 2-year rolling averages beginning in 1993.

SOURCE: Michael Planty and Jennifer L. Truman, "Figure 7. Nonfatal Firearm Violence, by Race and Hispanic Origin, 1994–2011," in *Firearm Violence, 1993–2011*, U.S. Department of Justice, Bureau of Justice Statistics, May 2013, http://www.bjs.gov/content/pub/pdf/fv9311.pdf (accessed July 16, 2014)

FIGURE 5.10

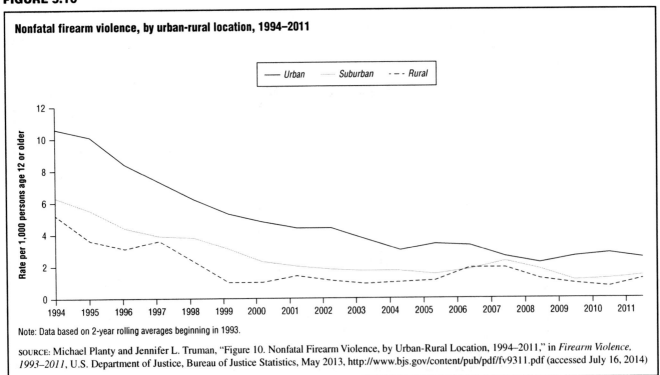

Nonfatal firearm violence, by urban-rural location, 1994–2011

Note: Data based on 2-year rolling averages beginning in 1993.

SOURCE: Michael Planty and Jennifer L. Truman, "Figure 10. Nonfatal Firearm Violence, by Urban-Rural Location, 1994–2011," in *Firearm Violence, 1993–2011*, U.S. Department of Justice, Bureau of Justice Statistics, May 2013, http://www.bjs.gov/content/pub/pdf/fv9311.pdf (accessed July 16, 2014)

Robbery and Aggravated Assault

Since 1993 firearms have been used in the overwhelming majority of homicides. (See Table 5.18.) This is not the case for other types of violent victimization.

Between 1993 and 2011 the proportion of nonfatal violence that involved firearms ranged from a high of 9.2% (in 1994) to a low of 5.8% (in 2008), most often remaining between 7% and 8%. The forms of nonfatal violence

TABLE 5.15

Nonfatal firearm violence, by population size, 1997–2011

	Rate per 1,000 persons age 12 or older					
Year	Not a place*	Less than 100,000	100,000– 249,999	250,000– 499,999	500,000– 999,999	1 million or more
1997	3.9	3.8	7.0	10.3	7.3	7.3
1998	3.0	3.9	4.8	7.0	9.2	5.7
1999	1.9	3.1	3.1	5.5	9.0	6.4
2000	1.5	2.2	3.9	6.5	6.3	5.6
2001	1.4	2.1	4.1	6.1	5.5	5.1
2002	1.2	2.3	2.8	3.9	4.9	5.3
2003	1.4	2.0	2.8	3.3	5.1	3.6
2004	1.4	1.4	3.0	4.1	5.5	2.7
2005	1.2	1.6	2.9	3.6	4.5	4.6
2006	1.6	2.1	2.6	2.6	3.8	4.9
2007	1.5	2.6	2.7	2.4	5.4	2.1
2008	0.8	2.1	2.1	3.2	4.9	1.4
2009	0.9	1.1	2.2	3.0	4.0	3.5
2010	0.9	1.2	1.8	2.8	5.1	4.0
2011	1.4	1.2	1.3	3.9	4.6	3.2

*A concentration of population that is not either legally bounded as an incorporated place having an active government or delineated for statistical purposes as a census designated place with definite geographic boundaries, such as a city, town, or village.
Note: Data based on 2-year rolling averages beginning in 1996. Population size information was not available from 1993 to 1995.

SOURCE: Michael Planty and Jennifer L. Truman, "Table 5. Nonfatal Firearm Violence, by Population Size, 1997–2011," in *Firearm Violence, 1993–2011*, U.S. Department of Justice, Bureau of Justice Statistics, May 2013, http://www.bjs.gov/content/pub/pdf/fv9311.pdf (accessed July 16, 2014)

TABLE 5.16

Fatal and nonfatal firearm violence, by age, 1993–2011

	Nonfatal firearm violence rate per 1,000 persons age 12 or older				
Year	12–17	18–24	25–34	35–49	50 or older
1993	—	—	—	—	—
1994	11.4	18.1	8.7	6.3	1.6
1995	9.8	16.1	7.7	5.5	1.6
1996	7.6	12.3	6.8	4.8	1.4
1997	7.1	12.8	5.4	4.5	1.2
1998	5.7	12.4	4.5	3.8	1.0
1999	4.7	8.9	4.6	2.6	0.7
2000	3.2	7.0	3.6	2.5	1.0
2001	2.2	6.8	3.1	2.4	1.0
2002	2.4	7.3	3.1	1.8	0.8
2003	2.8	6.3	2.7	1.6	0.7
2004	1.9	3.9	2.5	2.1	0.8
2005	1.2	4.4	3.1	1.8	1.0
2006	2.3	5.6	3.4	1.8	1.0
2007	4.3	4.6	3.0	2.2	0.9
2008	3.5	3.2	2.7	1.6	0.7
2009	0.9	3.9	2.3	1.5	0.6
2010	0.6!	5.8	2.0	1.3	0.6
2011	1.4	5.2	2.2	1.4	0.7

—Not applicable.
!Interpret with caution. Estimate based on 10 or fewer sample cases or coefficient of variation is greater than 50%.
Note: Nonfatal firearm violence data based on 2-year rolling averages beginning in 1993. Homicide data are annual estimates.

SOURCE: Adapted from Michael Planty and Jennifer L. Truman, "Table 4. Fatal and Nonfatal Firearm Violence, by Age, 1993–2011," in *Firearm Violence, 1993–2011*, U.S. Department of Justice, Bureau of Justice Statistics, May 2013, http://www.bjs.gov/content/pub/pdf/fv9311.pdf (accessed July 16, 2014)

most likely to involve firearms are robbery and aggravated assault. In 2011 firearms were used in 25.7% of robberies and 30.6% of aggravated assaults, levels that represented an increase over the late 1990s and the first decade of the 21st century while being within the range of variation between 1993 and 2011 as a whole.

Among the offenses reported to law enforcement agencies about which weapon information was known, the proportion of robberies and aggravated assaults involving firearms differed somewhat. In 2013, 122,266 robberies, or 41.1% of the total reported to police, involved firearms. (See Table 5.19.) Strong-arm incidents, involving physical force but no other external weapon, accounted for a slightly higher proportion of robberies (126,341). Among aggravated assaults reported to law enforcement agencies in 2013, those involving firearms accounted for 139,931, or 22.1% of the total. This was higher than the number of aggravated assaults committed with knives or other cutting instruments (120,063) and lower than the number committed with hands, fists, feet, or other body parts (169,644) and with weapons other than firearms, knives, or body parts (205,112).

There were regional differences, as well, in the percentage distribution of firearm robberies and aggravated assaults among all robberies and aggravated assaults in 2013. Robberies involving firearms accounted for 49% of total robberies in the South, 44.4% in the Midwest, 31%

in the Northeast, and 30.8% in the West. (See Table 5.20.) Similarly, firearm aggravated assaults represented a higher proportion of total aggravated assaults in the South (25%) and Midwest (24.4%) than in the West (18.3%) or Northeast (14.5%). (See Table 5.21.)

Justifiable Homicide

The FBI also compiles data on the category of justifiable homicide, which it defines in *Crime in the United States, 2013* as "the killing of a felon by a peace officer in the line of duty" or as "the killing of a felon, during the commission of a felony, by a private citizen."

Table 5.22 shows the data for justifiable homicides committed by law enforcement officers between 2009 and 2013. Almost all such incidents involved firearms. A firearm was used in 411 (all but three) cases in which a felon was killed by a law enforcement officer in the line of duty in 2009, 396 (all but one) cases in 2010, 401 (all but three) cases in 2011, 423 (all but three) cases in 2012, and 458 (all but three) cases in 2013. Handguns were the most frequently used type of firearm in cases of justifiable homicide by law enforcement officers. In 2013 out of 458 cases in which a firearm was used, 332 (72.5%) involved a handgun.

Table 5.23 shows the data for justifiable homicides committed by private citizens between 2009 and 2013.

TABLE 5.17

Nonfatal firearm injuries, 2007–11

	Firearm violence	
Injury and treatment	Number	Percent
Victims	2,218,500	100%
Not injured	1,707,800	77.0
Injured	510,700	23.0
Serious[a]	148,300	6.7
Gun shot	46,000	2.1
Minor[b]	357,100	16.1
Rape without other injuries	5,400!	0.2!
Treatment for injury[c]	510,700	100%
No treatment	140,700	27.5
Any treatment	370,000	72.5
Treatment setting[d]	370,000	100%
At the scene/home of victim, neighbor, or friend/location	68,000	18.4
In doctor's office/hospital emergency room/ overnight at hospital	302,000	81.6

!Interpret with caution. Estimate based on 10 or fewer sample cases, or coefficient of variation is greater than 50%.
[a]Includes injuries such as gun shots, knife wounds, internal injuries, unconsciousness, and broken bones.
[b]Includes bruises, cuts, and other minor injuries.
[c]Includes only victims who were injured.
[d]Includes only victims who were injured and received treatment.

SOURCE: Adapted from Michael Planty and Jennifer L. Truman, "Table 9. Nonfatal Firearm and Nonfirearm Violence, by Injury and Treatment Received, 2007–2011, in *Firearm Violence, 1993–2011*, U.S. Department of Justice, Bureau of Justice Statistics, May 2013, http://www.bjs.gov/content/pub/pdf/fv9311.pdf (accessed July 16, 2014)

TABLE 5.18

Percentage of violence involving a firearm, by type of crime, 1993–2011

Year	Homicide	Nonfatal violence[a]	Robbery	Aggravated assault
1993	71.2%	9.1%	22.3%	30.7%
1994	71.4	9.2	27.1	31.9
1995	69.0	7.8	27.3	28.0
1996	68.0	7.8	24.6	25.7
1997	68.0	7.6	19.9	27.0
1998	65.9	7.0	20.1	26.5
1999	64.1	6.0	19.2	22.4
2000	64.4	7.2	21.1	26.6
2001[b]	55.9	7.5	29.5	26.0
2002	67.1	7.3	23.4	28.7
2003	67.2	6.1	22.4	22.2
2004	67.0	6.8	19.7	23.6
2005	68.2	7.2	21.8	25.7
2006	68.9	7.3	16.6	24.3
2007	68.8	8.1	20.0	32.6
2008	68.3	5.8	19.6	24.6
2009	68.4	7.2	27.0	23.2
2010	68.1	8.4	24.7	25.4
2011[c]	69.6	8.0	25.7	30.6

[a]Nonfatal violence includes rape, sexual assault, robbery, aggravated and simple assault. A small percentage of rape and sexual assaults involved firearms but are not shown in table due to small sample sizes.
[b]The homicide estimates that occurred as a result of the events of September 11, 2001, are included in the total number of homicides.
[c]Preliminary homicide estimates retrieved from Hoyert DL, Xu JQ. (2012) Deaths: Preliminary data for 2011. *National Vital Statistics Reports*, 61(6).

SOURCE: Michael Planty and Jennifer L. Truman, "Table 2. Percent of Violence Involving a Firearm, by Type of Crime," in *Firearm Violence, 1993–2011*, U.S. Department of Justice, Bureau of Justice Statistics, May 2013, http://www.bjs.gov/content/pub/pdf/fv9311.pdf (accessed July 16, 2014)

Firearms were used in the overwhelming majority of cases, but their use in justifiable homicides by citizens was not as universal as in justifiable homicides by law enforcement officers. Firearms were used in 218 of 266 (82%) cases of justifiable homicide by private citizens in 2009 and in 223 of 281 (79.4%) cases in 2013. As with cases involving law enforcement officers, handguns were the most frequently used type of firearm in cases of justifiable homicide by private citizens. Out of 223 cases in which a firearm was used in 2013, 171 (76.7%) involved a handgun.

WHERE DO CRIMINALS GET FIREARMS?

Many criminals obtain their guns legally. For example, James Holmes (1987–), accused of killing 12 people and wounding 58 others in the Aurora, Colorado, movie theater shooting in July 2012, was armed with an AR-15 assault rifle, a Remington 12-gauge shotgun, and two Glock handguns at the time of the incident. His four weapons had been purchased at shops in the Denver area, and he bought a supply of ammunition online. In Colorado no permit or waiting period is required for gun purchases. However, firearms dealers are required by law to maintain records on all sales, including personal information about the purchaser, the serial number and identifying information about the gun, and the date and terms of the sale. Speaking at a press conference covered by ABC News on July 20, 2012 (http://abcnews.go.com/US/colorado-movie-theater-shooting-suspect-bought-guns-6000/story?id=16817842#.UHbBQa4fjd4), the Aurora chief of police Dan Oates said of Holmes, "All the ammunition he possessed, he possessed legally, all the weapons he possessed, he possessed legally, all the clips he possessed, he possessed legally."

Guns can also be obtained illegally in a variety of ways, primarily by stealing them or by illegally buying or trading them on the black market (a market where products are bought and sold illegally). However, there are few up-to-date studies of the relationship between criminality and gun ownership. This is partly a result of the Tiahrt Amendment (discussed in greater detail in Chapter 2), which began preventing the dissemination of certain statistics on the sales of multiple handguns and on firearms tracing statistics (gun trace statistics) in 2003. Additionally, federal funding for studies that might be used to support increased gun control has been staunchly opposed by the National Rifle Association of America and other gun rights advocates. In *Access Denied: How the Gun Lobby Is Depriving Police, Policy Makers, and the Public of the Data We Need to Prevent Gun Violence* (January 2013, http://3gbwir1ummda 16xrhf4do9d21bsx.wpengine.netdna-cdn.com/wp-content/uploads/2014/04/AccessDenied.pdf), Mayors against Illegal

TABLE 5.19

Robberies and aggravated assaults, by type of weapon used, 2012–13

	Robbery				Aggravated assault			
	Firearm	Knife or cutting instrument	Other weapon	Strong-arm	Firearm	Knife or cutting instrument	Other weapon	Hands, fists, feet, etc.
Total all agencies:								
2012	125,366	23,532	26,784	129,662	146,045	125,595	218,469	180,637
2013	122,266	22,553	26,448	126,341	139,931	120,063	205,112	169,644
Percent change	−2.5	−4.2	−1.3	−2.6	−4.2	−4.4	−6.1	−6.1

SOURCE: Adapted from "Table 15. Crime Trends, Additional Information about Selected Offenses by Population Group, 2012–2013," in *Crime in the United States, 2013*, Federal Bureau of Investigation, 2014, http://www.fbi.gov/about-us/cjis/ucr/crime-in-the-u.s/2013/crime-in-the-u.s.-2013/tables/table-15/table_15_crime_trends_by_additional_information_about_selected_offenses_2012-2013.xls (accessed November 20, 2014)

TABLE 5.20

Robbery, by type of weapon used, percentage distribution by region, 2013

		Armed			
Region	Total all weapons*	Firearms	Knives or cutting instruments	Other weapons	Strong-arm
Total	**100**	**40.0**	**7.6**	**8.9**	**43.6**
Northeast	100	31.0	9.1	7.9	51.9
Midwest	100	44.4	5.5	9.1	41.0
South	100	49.0	6.6	8.5	35.9
West	100	30.8	9.0	10.1	50.1

*Because of rounding, the percentages may not add to 100.0.

SOURCE: "Robbery Table 3. Robbery, Types of Weapons Used, Percent Distribution by Region, 2013," in *Crime in the United States, 2013*, Federal Bureau of Investigation, 2014, http://www.fbi.gov/about-us/cjis/ucr/crime-in-the-u.s/2013/crime-in-the-u.s.-2013/violent-crime/robbery-topic-page/robbery_table_3_robbery_types_of_weapons_used_percent_distribution_by_region_2013.xls (accessed November 20, 2014)

TABLE 5.21

Aggravated assault, by type of weapon used, percentage distribution by region, 2013

Region	Total all weapons*	Firearms	Knives or cutting instruments	Other weapons (clubs, blunt objects, etc.)	Personal weapons (hands, fists, feet, etc.)
Total	**100**	**21.6**	**19.1**	**32.2**	**27.0**
Northeast	100	14.5	22.9	31.0	31.7
Midwest	100	24.4	16.8	29.8	28.9
South	100	25.0	19.5	33.0	22.5
West	100	18.3	17.6	33.2	30.9

*Because of rounding, the percentages may not add to 100.0.

SOURCE: "Aggravated Assault Table. Aggravated Assault, Types of Weapons Used, Percent Distribution by Region, 2013," in *Crime in the United States, 2013*, Federal Bureau of Investigation, 2014, http://www.fbi.gov/about-us/cjis/ucr/crime-in-the-u.s/2013/crime-in-the-u.s.-2013/violent-crime/aggravated-assault-topic-page/aggravated_assault_table_aggravated_assault_types_of_weapons_used_percent_distribution_by_region_2013.xls (accessed November 20, 2014)

Guns (MAIG), a gun control advocacy group, explains that pressure from the gun rights lobby has led the National Institute of Justice (NIJ), a primary research branch of the DOJ, to defund much of the firearms research that was once a top priority. The MAIG notes that although the NIJ funded 32 studies related to firearms between 1993 and 1999, "it has not funded a single public study on firearms during the Obama Administration."

Thus, the most reliable studies on the topic date from the 1990s and the first decade of the 21st century. The most comprehensive study conducted by the government on where criminals get their guns was released in 2001 and was based on interviews with 18,000 state prison inmates. In *Firearm Use by Offenders* (November 2001, http://bjs.ojp.usdoj.gov/content/pub/pdf/fuo.pdf), Caroline Wolf Harlow of the BJS indicates that 13.9% of those

TABLE 5.22

Justifiable homicides committed by law enforcement officers, by weapon used, 2009–13*

Year	Total	Total firearms	Handguns	Rifles	Shotguns	Firearms, type not stated	Knives or cutting instruments	Other dangerous weapons	Personal weapons
2009	414	411	326	29	6	50	0	3	0
2010	397	396	323	29	6	38	0	1	0
2011	404	401	305	36	11	49	2	0	1
2012	426	423	339	38	7	39	0	3	0
2013	461	458	332	46	9	71	0	3	0

*The killing of a felon by a law enforcement officer in the line of duty.

SOURCE: "Expanded Homicide Data Table 14. Justifiable Homicide, by Weapon, Law Enforcement, 2009–2013," in *Crime in the United States, 2013*, Federal Bureau of Investigation, 2014, http://www.fbi.gov/about-us/cjis/ucr/crime-in-the-u.s/2013/crime-in-the-u.s.-2013/offenses-known-to-law-enforcement/expanded-homicide/expanded_homicide_data_table_14_justifiable_homicide_by_weapon_law_enforcement_2009-2013.xls (accessed November 20, 2014)

TABLE 5.23

Justifiable homicides committed by private citizens, by weapon used, 2009–13*

Year	Total	Total firearms	Handguns	Rifles	Shotguns	Firearms, type not stated	Knives or cutting instruments	Other dangerous weapons	Personal weapons
2009	266	218	167	9	19	23	30	10	8
2010	285	236	170	8	30	28	33	11	5
2011	270	209	156	13	11	29	49	9	3
2012	315	263	198	20	15	30	35	6	11
2013	281	223	171	5	12	35	35	13	10

*The killing of a felon, during the commission of a felony, by a private citizen.

SOURCE: "Expanded Homicide Data Table 15. Justifiable Homicide, by Weapon, Private Citizen, 2009–2013," in *Crime in the United States, 2013*, Federal Bureau of Investigation, 2014, http://www.fbi.gov/about-us/cjis/ucr/crime-in-the-u.s/2013/crime-in-the-u.s.-2013/offenses-known-to-law-enforcement/expanded-homicide/expanded_homicide_data_table_15_justifiable_homicide_by_weapon_private_citizen_2009-2013.xls (accessed November 20, 2014)

who carried a firearm during the offense for which they were serving time in 1997 bought their gun from a retail store, pawn shop, flea market, or gun show. This figure was down from 20.8% in 1991, when the previous survey was conducted. Another 39.6% acquired their firearms from family or friends, up from 33.8% in 1991. The remaining 39.2% acquired their firearms "on the street" from an illegal source, down from 40.8% in 1991.

The MAIG attempted to fill the research gap by hiring an investigative services firm to determine where criminals get their guns. The results of this investigation were published in *Inside Straw Purchasing: How Criminals Get Guns Illegally* (April 2008). The coalition presents findings on straw purchasing, a situation in which a person who would be denied a gun purchase, such as a convicted felon or an underaged buyer, has another person (the straw purchaser) fill out the paperwork and obtain the gun for him or her. Sometimes a straw purchaser is used when a person simply does not want a gun purchase listed in his or her name.

The MAIG explains that straw purchasing occurred frequently in so-called easy stores. Straw purchasers

often paid for their purchases with both money and drugs and bought several guns at one time. Gun dealers appear to be the key in this illegal activity: some encourage straw purchases and actually coach straw purchasers, whereas others discourage the practice by asking many probing questions until the straw purchaser leaves the store. These dealers also train their employees how to spot straw purchasers and thwart them.

Many criminals get their guns from the black market. Sometimes, these guns are legally purchased in states with less restrictive gun laws and are then transported to states with strong gun laws, a phenomenon known as gunrunning or gun trafficking (buying, moving, and selling guns illegally). In *Trace the Guns: The Link between Gun Laws and Interstate Gun Trafficking* (September 2010, http://every town.org/documents/2014/10/trace-the-guns.pdf), the MAIG analyzes data related to guns recovered from crime scenes in 2009 and gun laws in the states in which the guns were purchased, finding that states with weak gun laws were more likely to serve as a source state in supplying guns to criminals. The coalition reports that the Bureau of Alcohol, Tobacco, Firearms, and Explosives (ATF) was able to identify the source state of 145,321 (61%) out of

238,107 guns that were recovered at U.S. crime scenes in 2009. Of this number, 70% (102,067) of the guns were used in crimes in the same state in which they were purchased, and 30% (43,254) of the traced guns were recovered in a different state. Nearly half (20,996 firearms, or 48.5%) of the traced guns that were used in crimes in other states in 2009 originated in just 10 states: Georgia (2,781), Florida (2,640), Virginia (2,557), Texas (2,240), Indiana (2,011), Ohio (1,806), Pennsylvania (1,777), North Carolina (1,775), California (1,772), and Arizona (1,637). In 2009 the states that had the highest rates of guns recovered at crime scenes in other states (i.e., the number of guns that were used in crimes in other states per 100,000 inhabitants) were: Mississippi (50.3), West Virginia (46.8), Kentucky (34.9), Alaska (33.4), Alabama (33.2), South Carolina (33), Virginia (32.4), Indiana (31.3), Nevada (30.6), and Georgia (28.3). All of these states had rates that were more than twice the national average of 14.1 per 100,000 inhabitants.

In *Trace the Guns*, the MAIG also analyzes "Time-to-Crime" (TTC) data (the time between the initial sale of a gun and the time it is recovered at a crime scene) to determine the likelihood that the gun was acquired illegally before being used in a crime. Firearms that have a TTC of less than two years are considered most likely to have been trafficked illegally. The national average TTC in 2009 was 10.8 years. The MAIG reports that the 10 states with the highest gun export rates also had a higher proportion of guns recovered in other states in less than two years from the initial retail sale.

In addition, the MAIG finds that states that had enacted strong gun laws were less likely to serve as a source for interstate trafficking of firearms in 2009. When used together, state regulations that require the prosecution of straw purchasers or those who falsify purchase information, stipulate background checks for handgun sales at gun shows, require purchase permits, authorize local officials to approve or deny concealed carry permits, deny guns to people convicted of violent misdemeanors, require owners to report missing guns to law enforcement, support local firearms regulations, and require state oversight of gun dealers, seemed to offer some effectiveness in limiting illegal trafficking. The MAIG finds that the states that had not enacted these laws were twice as likely to serve as a source state for guns used in crimes in other states and for guns with a short TTC.

Operation Fast and Furious

In a case that drew attention to the role of the U.S. government in supplying guns to criminals, ATF agents allowed an estimated 2,200 firearms to be trafficked illegally from U.S. border states into the arsenals of Mexican narcotics cartels. Beginning in October 2009, agents in what was known as Operation Fast and Furious

were instructed by their superiors not to arrest those on the lowest end of the crime organization, the so-called straw purchasers, who falsified information required to purchase guns, but rather to allow the guns to be traded on the black market in the hope that U.S. agents could infiltrate the notoriously violent Mexican crime groups at a higher level of leadership. Sari Horwitz explains in "Operation Fast and Furious: A gunrunning sting gone wrong" (WashingtonPost.com, July 26, 2011) that "in drug-trafficking cases, investigating agents, by law, cannot let drugs 'walk' onto the street. Since gun sales are legal, agents on surveillance are not required to step in and stop weapons from hitting the streets." However, a group of agents in Phoenix, Arizona, became increasingly alarmed about the potential negative effects of the operation, and when U.S. Border Patrol Agent Brian Terry was killed in a firefight in December 2010 with a gang using an AK-47 assault rifle that had been trafficked to Mexico through Fast and Furious connections, they made the operation public and a congressional investigation ensued. Horwitz notes that of the 2,200 firearms known to have been trafficked in the case, 227 had been recovered in connection with crimes in Mexico, 363 had been recovered in U.S. crime cases, and 1,430 remained "on the streets" as of July 2011.

In November 2012 Michael E. Horowitz, the U.S. inspector general, released the report *A Review of ATF's Operation Fast and Furious and Related Matters* (http://www.justice.gov/oig/reports/2012/s1209.pdf), the result of a review of more than 100,000 documents and interviews with over 100 individuals connected with the case. Horowitz finds "a series of misguided strategies, tactics, errors in judgment, and management failures that permeated ATF Headquarters and the Phoenix Field Division, as well as the U.S. Attorney's Office for the District of Arizona." Although disciplinary action was recommended for some personnel in the operation, no recommendations for criminal prosecution were made. Furthermore, the U.S. attorney general Eric Holder (1951–), who had been a key target of Republican critics of the case, had not been fully advised of the operation and was thus cleared of responsibility.

WEAPONS OFFENSES

Weapons offenses are violations of statutes or regulations that control deadly weapons, which include firearms and their ammunition, silencers, explosives, and certain knives. All 50 states, many cities and towns, and the federal government have laws concerning deadly weapons, including restrictions on their possession, carrying, use, sales, manufacturing, importing, and exporting.

Table 5.24 shows arrest trends for crimes committed in 2004 and in 2013, based on reports from 7,858 law enforcement agencies. Total arrests across all age groups

TABLE 5.24

Arrest trends, by offense and age, 2004–13

[7,858 agencies; 2013 estimated population 192,473,854; 2004 estimated population 178,805,123]

	Total all ages			Under 18 years of age			18 years of age and over		
Offense charged	2004	2013	Percent change	2004	2013	Percent change	2004	2013	Percent change
Total[a]	8,402,488	7,120,525	−15.3	1,226,865	666,263	−45.7	7,175,623	6,454,262	−10.1
Murder and nonnegligent manslaughter	7,872	6,695	−15.0	643	492	−23.5	7,229	6,203	−14.2
Rape[b]	15,019	10,471	−30.3	2,414	1,484	−38.5	12,605	8,987	−28.7
Robbery	64,349	61,019	−5.2	14,936	12,340	−17.4	49,413	48,679	−1.5
Aggravated assault	270,826	234,554	−13.4	35,912	19,351	−46.1	234,914	215,203	−8.4
Burglary	180,617	163,261	−9.6	49,721	27,960	−43.8	130,896	135,301	3.4
Larceny-theft	721,769	771,869	6.9	198,071	117,141	−40.9	523,698	654,728	25.0
Motor vehicle theft	87,337	41,385	−52.6	22,784	7,367	−67.7	64,553	34,018	−47.3
Arson	9,093	6,792	−25.3	4,593	2,370	−48.4	4,500	4,422	−1.7
Violent crime[c]	358,066	312,739	−12.7	53,905	33,667	−37.5	304,161	279,072	−8.2
Property crime[c]	998,816	983,307	−1.6	275,169	154,838	−43.7	723,647	828,469	14.5
Other assaults	769,970	696,659	−9.5	148,743	91,436	−38.5	621,227	605,223	−2.6
Forgery and counterfeiting	71,993	37,884	−47.4	2,988	649	−78.3	69,005	37,235	−46.0
Fraud	196,788	88,245	−55.2	4,622	2,755	−40.4	192,166	85,490	−55.5
Embezzlement	11,995	10,202	−14.9	698	233	−66.6	11,297	9,969	−11.8
Stolen property; buying, receiving, possessing	78,027	58,443	−25.1	13,879	6,354	−54.2	64,148	52,089	−18.8
Vandalism	160,941	128,589	−20.1	61,262	29,676	−51.6	99,679	98,913	−0.8
Weapons; carrying, possessing, etc.	110,697	91,150	−17.7	25,478	12,771	−49.9	85,219	78,379	−8.0
Prostitution and commercialized vice	55,369	35,562	−35.8	1,157	550	−52.5	54,212	35,012	−35.4
Sex offenses (except rape and prostitution)	54,292	35,604	−34.4	10,923	6,249	−42.8	43,369	29,355	−32.3
Drug abuse violations	1,080,301	976,882	−9.6	118,392	75,767	−36.0	961,909	901,115	−6.3
Gambling	6,365	4,400	−30.9	1,099	569	−48.2	5,266	3,831	−27.3
Offenses against the family and children	73,249	60,479	−17.4	3,608	1,668	−53.8	69,641	58,811	−15.6
Driving under the influence	840,325	710,351	−15.5	11,212	4,315	−61.5	829,113	706,036	−14.8
Liquor laws	343,782	206,285	−40.0	71,976	34,283	−52.4	271,806	172,002	−36.7
Drunkenness	363,978	300,708	−17.4	11,117	5,107	−54.1	352,861	295,601	−16.2
Disorderly conduct	364,859	258,950	−29.0	110,374	53,471	−51.6	254,485	205,479	−19.3
Vagrancy	23,137	18,154	−21.5	3,138	545	−82.6	19,999	17,609	−12.0
All other offenses (except traffic)	2,372,952	2,080,548	−12.3	230,539	125,976	−45.4	2,142,413	1,954,572	−8.8
Suspicion	4,249	399	−90.6	535	38	−92.9	3,714	361	−90.3
Curfew and loitering law violations	66,586	25,384	−61.9	66,586	25,384	−61.9	—	—	—

[a]Does not include suspicion.
[b]The rape figures in this table are based on the legacy definition of rape only. The rape figures shown include converted National Incident-Based Reporting System rape data and those states/agencies that reported the legacy definition of rape for both years.
[c]Violent crimes in this table are offenses of murder and nonnegligent manslaughter, rape (legacy definition), robbery, and aggravated assault. Property crimes are offenses of burglary, larceny-theft, motor vehicle theft, and arson.

SOURCE: "Table 32. Ten-Year Arrest Trends, Totals, 2004–2013," in *Crime in the United States, 2013*, Federal Bureau of Investigation, 2014, http://www.fbi.gov/about-us/cjis/ucr/crime-in-the-u.s/2013/crime-in-the-u.s.-2013/tables/table-32/table_32_ten_year_arrest_trends_totals_2013.xls (accessed November 20, 2014)

decreased 15.3% during this 10-year period, and arrests for weapons offenses across all age groups decreased 17.7%. Arrests of people under the age of 18 years fell 45.7% during this period, and arrests of those under the age of 18 years for weapons offenses fell 49.9%. By comparison, arrests of those aged 18 years and older fell 10.1%, and arrests of those aged 18 years and older on weapons offenses fell 8%.

Table 5.25 shows arrests by race and age for crimes committed in 2013. (The data in Table 5.25, which were drawn from a larger number of law enforcement agencies [11,951], differ slightly from the data in Table 5.24.) White offenders accounted for 68.9% of total arrests in 2013, and African American offenders accounted for 28.3%. Whites accounted for 58.2% of weapons offenses and African Americans accounted for 39.8%. In both categories of arrests, as well as in many specific categories, African Americans were overrepresented. According to the U.S. Census Bureau (June 2014, http://factfinder2.census.gov/faces/nav/jsf/pages/index.xhtml), African Americans accounted for approximately 41.6 million, or 13.2%, of the United States' total population of 316.1 million in 2013.

TABLE 5.25

Arrests, by race, 2013

[11,951 agencies; 2013 estimated population 245,741,701]

Offense charged	Total arrests — Race						Percent distribution[a] — Race						Total arrests — Ethnicity			Percent distribution[a] — Ethnicity		
	Total	White	Black or African American	American Indian or Alaska Native	Asian	Native Hawaiian or other Pacific Islander	Total	White	Black or African American	American Indian or Alaska Native	Asian	Native Hawaiian or other Pacific Islander	Total[b]	Hispanic or Latino	Not Hispanic or Latino	Total	Hispanic or Latino	Not Hispanic or Latino
Total	**9,014,635**	**6,214,197**	**2,549,655**	**140,290**	**105,109**	**5,384**	**100**	**68.9**	**28.3**	**1.6**	**1.2**	**0.1**	**4,813,531**	**799,931**	**4,013,600**	**100**	**16.6**	**83.4**
Murder and nonnegligent manslaughter	8,383	3,799	4,379	98	101	6	100	45.3	52.2	1.2	1.2	0.1	4,850	1,052	3,798	100	21.7	78.3
Rape[c]	13,515	8,946	4,229	160	173	7	100	66.2	31.3	1.2	1.3	0.1	9,671	2,006	7,665	100	20.7	79.3
Robbery	78,538	32,945	44,271	579	649	94	100	41.9	56.4	0.7	0.8	0.1	43,475	8,481	34,994	100	19.5	80.5
Aggravated assault	291,031	183,092	98,748	4,356	4,423	412	100	62.9	33.9	1.5	1.5	0.1	178,817	43,604	135,213	100	24.4	75.6
Burglary	203,089	136,990	61,709	1,966	2,196	228	100	67.5	30.4	1.0	1.1	0.1	120,667	24,035	96,632	100	19.9	80.1
Larceny-theft	990,936	677,173	284,358	16,402	12,605	398	100	68.3	28.7	1.7	1.3	*	495,161	62,954	432,207	100	12.7	87.3
Motor vehicle theft	52,307	34,864	15,960	685	725	73	100	66.7	30.5	1.3	1.4	0.1	31,455	8,203	23,252	100	26.1	73.9
Arson	8,364	6,198	1,925	130	107	4	100	74.1	23.0	1.6	1.3	*	4,131	661	3,470	100	16.0	84.0
Violent crime[d]	391,467	228,782	151,627	5,193	5,346	519	100	58.4	38.7	1.3	1.4	0.1	236,813	55,143	181,670	100	23.3	76.7
Property crime[d]	1,254,696	855,225	363,952	19,183	15,633	703	100	68.2	29.0	1.5	1.2	0.1	651,414	95,853	555,561	100	14.7	85.3
Other assaults	881,086	573,546	283,357	14,041	9,717	425	100	65.1	32.2	1.6	1.1	*	453,745	65,939	387,806	100	14.5	85.5
Forgery and counterfeiting	48,581	31,208	16,375	288	677	33	100	64.2	33.7	0.6	1.4	0.1	26,000	3,864	22,136	100	14.9	85.1
Fraud	112,920	74,682	35,958	1,145	1,094	41	100	66.1	31.8	1.0	1.0	*	58,355	5,323	53,032	100	9.1	90.9
Embezzlement	12,574	7,882	4,386	87	207	12	100	62.7	34.9	0.7	1.6	0.1	7,380	721	6,659	100	9.8	90.2
Stolen property; buying, receiving, possessing	74,541	50,237	22,687	684	862	71	100	67.4	30.4	0.9	1.2	0.1	43,512	9,990	33,522	100	23.0	77.0
Vandalism	161,078	113,842	42,566	2,951	1,638	81	100	70.7	26.4	1.8	1.0	0.1	83,109	15,173	67,936	100	18.3	81.7
Weapons; carrying, possessing, etc.	112,228	65,317	44,671	888	1,251	101	100	58.2	39.8	0.8	1.1	0.1	59,587	15,282	44,305	100	25.6	74.4
Prostitution and commercialized vice	41,946	22,666	17,378	386	1,492	24	100	54.0	41.4	0.9	3.6	0.1	21,510	3,801	17,709	100	17.7	82.3
Sex offenses (except rape and prostitution)	46,553	33,695	11,462	622	744	30	100	72.4	24.6	1.3	1.6	0.1	22,165	6,065	16,100	100	27.4	72.6
Drug abuse violations	1,204,162	815,181	365,785	9,408	12,930	858	100	67.7	30.4	0.8	1.1	0.1	628,248	118,827	509,421	100	18.9	81.1
Gambling	5,055	1,433	3,362	27	226	7	100	28.3	66.5	0.5	4.5	0.1	1,451	276	1,175	100	19.0	81.0
Offenses against the family and children	78,465	51,017	25,519	1,414	511	4	100	65.0	32.5	1.8	0.7	*	40,416	2,509	37,907	100	6.2	93.8
Driving under the influence	910,470	766,440	113,928	12,575	16,831	696	100	84.2	12.5	1.4	1.8	0.1	505,788	106,377	399,411	100	21.0	79.0
Liquor laws	277,444	222,201	40,665	10,861	3,672	45	100	80.1	14.7	3.9	1.3	*	147,418	18,692	128,726	100	12.7	87.3
Drunkenness	356,427	288,146	56,885	7,399	3,550	447	100	80.8	16.0	2.1	1.0	0.1	254,208	44,007	210,201	100	17.3	82.7
Disorderly conduct	372,202	231,604	129,782	7,982	2,775	59	100	62.2	34.9	2.1	0.7	*	191,115	20,110	171,005	100	10.5	89.5
Vagrancy	21,354	13,732	6,802	581	222	17	100	64.3	31.9	2.7	1.0	0.1	11,970	1,728	10,242	100	14.4	85.6
All other offenses (except traffic)	2,602,939	1,741,855	790,854	43,953	25,090	1,187	100	66.9	30.4	1.7	1.0	*	1,337,075	204,652	1,132,423	100	15.3	84.7
Suspicion	825	499	303	12	11	0	100	60.5	36.7	1.5	1.3	0.0	130	6	124	100	4.6	95.4
Curfew and loitering law violations	47,622	25,007	21,351	610	630	24	100	52.5	44.8	1.3	1.3	0.1	32,122	5,593	26,529	100	17.4	82.6

*Less than one-tenth of 1 percent.

[a]Because of rounding, the percentages may not add to 100.0.

[b]The ethnicity totals are representative of those agencies that provided ethnicity breakdowns. Not all agencies provide ethnicity data; therefore, the race and ethnicity totals will not be equal.

[c]The rape figures in this table are an aggregate total of the data submitted using both the revised and legacy Uniform Crime Reporting definitions.

[d]Violent crimes in this table are offenses of murder and nonnegligent manslaughter, rape (revised and legacy definitions), robbery, and aggravated assault. Property crimes are offenses of burglary, larceny-theft, motor vehicle theft, and arson.

SOURCE: "Table 43A. Arrests, by Race, 2013," in *Crime in the United States, 2013*, Federal Bureau of Investigation, 2014, http://www.fbi.gov/about-us/cjis/ucr/crime-in-the-u.s/2013/crime-in-the-u.s.-2013/tables/table-43 (accessed November 20, 2014)

CHAPTER 6

GUN-RELATED INJURIES AND FATALITIES

The public health community, which is represented at the national level by the Centers for Disease Control and Prevention (CDC), believes that collecting comprehensive data on firearm injuries and deaths—such as who was shot, under what circumstances, and with what kind of weapon—is the first step in reducing these injuries and deaths. The next step, from a public health approach, would logically be a campaign similar to those that eradicated polio and reduced traffic fatalities.

However, the CDC's ability to treat firearm injuries as a public health problem is limited. Christine Jamieson explains in "Gun Violence Research: History of the Federal Funding Freeze" (February 2013, http://www.apa.org/ science/about/psa/2013/02/gun-violence.aspx) that the CDC's ability to research firearm violence has been tightly restricted since 1996. During the mid-1990s, after having funded research demonstrating that people who lived in gun-owning households were at a greater risk of homicide than those who lived in non-gun households, the CDC came under attack from the National Rifle Association of America (NRA) and other gun rights groups. As a consequence of an NRA campaign to prohibit federally funded research that might lead to increased gun control, Congress passed legislation in 1996 stating that "none of the funds made available for injury prevention and control at the Centers for Disease Control and Prevention may be used to advocate or promote gun control." Congress also reappropriated the exact amount of money that the CDC had previously used to fund firearms injury research ($2.6 million) for traumatic brain injury research. According to Mayors against Illegal Guns, in *Access Denied: How the Gun Lobby Is Depriving Police, Policy Makers, and the Public of the Data We Need to Prevent Gun Violence* (January 2013, http://3gbwir1ummda16xrhf4do9d21bsx .wpengine.netdna-cdn.com/wp-content/uploads/2014/04/ AccessDenied.pdf), as of 2012 only $100,000 of the CDC's $5.6 billion budget was allocated to research into firearm injury prevention.

Nevertheless, the CDC supplies much of the basic, authoritative data regarding victims of firearm violence. Its National Center for Injury Prevention and Control administers a system that tracks the numbers of firearm-related injuries, and these data are made available through the Web-Based Injury Statistics Query and Reporting System (WISQARS; http://www.cdc.gov/ injury/wisqars/index.html). This interactive database provides reports of injury-related data of all types, including firearm injuries, both fatal and nonfatal.

The CDC also has a state-based system, the National Violent Death Reporting System (http://www.cdc.gov/ ViolencePrevention/NVDRS/index.html), that collects information about violent deaths, including firearm deaths, from a variety of sources in some states. These sources include law enforcement, medical examiners and coroners, crime laboratories, and death certificates. These data help detail the circumstances that might have contributed to the firearm death. Furthermore, they help each participating state design and implement prevention and intervention efforts tailored to that state's needs.

NONFATAL GUNSHOT INJURIES

Table 6.1 shows WISQARS data for the numbers of nonfatal gunshot injuries and the rates per 100,000 population between 2003 and 2012. Both the number and rate of nonfatal gunshot injuries were higher in 2012 than in 2003, but there was considerable fluctuation in the interim, and no clear trend was discernible. In any event, the CDC indicates in "Ten Leading Causes of Death and Injury" (2014, http://www.cdc.gov/injury/wisqars/LeadingCauses .html) that gunshot injuries were not among the 10 leading causes of nonfatal injury for people of any age who were treated in emergency departments between 2003 and 2012. Table 6.2 shows the 10 leading causes of nonfatal injuries for all age groups in 2012. Although there was variation among different age groups regarding the most common

TABLE 6.1

Nonfatal gunshot injuries and rates per 100,000 people, 2003–12

Year	Injuries	Population	Age-adjusted rate
2003	46,894	290,107,933	15.88
2004	47,834	292,805,298	16.15
2005	54,437	295,516,599	18.33
2006	56,739	298,379,912	18.74
2007	54,165	301,231,207	17.91
2008	61,406	304,093,966	20.11
2009	48,158	306,771,529	15.68
2010	59,344	308,745,538	19.35
2011	59,208	311,587,816	18.98
2012	64,034	313,914,040	20.42

SOURCE: Adapted from "Nonfatal Injury Reports, 2001–2012," in *Web-Based Injury Statistics Query and Reporting System (WISQARS)*, Centers for Disease Control and Prevention, 2014, http://webappa.cdc.gov/sasweb/ncipc/nfirates2001.html (accessed July 16, 2014)

causes of nonfatal injury in 2012, the most common injury causes for all age groups included falls, being struck by or against someone or something, and overexertion.

Table 6.3 shows nonfatal firearm injury numbers and rates among children between the ages of zero and 19 years in 2012. Both the numbers and the rates per 100,000 of nonfatal firearm injuries were low across all age groups, except for males aged 15 to 19 years. Of the nearly 11 million males in this age group, 8,493 were injured by firearms, for a rate of 77.5 per 100,000. This rate was nearly nine times higher than that of females in the same age group, who were the next most likely cohort to be injured by a firearm, and it was over 24 times higher than the rate for the next youngest cohort of males (aged 10 to 14 years). Overall, children were less likely to be injured by a BB or pellet gun. (See Table 6.4.) As with real firearms, the rate of injury from BB or pellet guns was highest for males aged 15 to 19 years, but the rate for that group, at 6.7 per 100,000, was more than 11 times lower than that cohort's firearm injury rate.

FIREARM FATALITIES

Figure 6.1 shows the trend in firearm homicide numbers and rates between 1993 and 2011. As noted in Chapter 5, firearm homicides decreased dramatically, both in raw numbers and in rates per 100,000 people, after peaking during the early 1990s. The number of firearm homicides nationwide in 2011, 11,101, represented a 39% decline from the 1993 peak of 18,253 firearm homicides. The number of firearm homicides reached its lowest point in 1999, at 10,828, and then rose to 12,791 in 2006, before declining to the 2011 level.

Table 6.5 provides a demographic overview of firearm deaths not only from homicides but also from suicides and unintentional shootings. This overall firearm-related death rate is broken down by sex, race, Hispanic origin, and age, as well as by date for selected years between 1970 and 2010. The overall age-adjusted death rate from firearm-related deaths remained somewhat stable between 1970 and 1990 at about 14 deaths per 100,000 population. The rate then dropped to 13.4 deaths per 100,000 population in 1995 and continued to drop to 10.2 deaths per 100,000 population in 2000. Thereafter, it stabilized at about 10 deaths per 100,000 population.

Nevertheless, this overall rate disguises drastic fluctuations by demographic characteristics. Females have consistently had a much lower firearm-related death rate than males. Although the age-adjusted female firearm-related death rate fell roughly in tandem with the male rate, it was at all times approximately five to six times lower than the age-adjusted rate for males. (See Table 6.5.) The age-adjusted rate for females ranged from a high of 4.8 per 100,000 in 1970 to a low of 2.7 in 2005 and 2010, whereas the age-adjusted rate for males ranged from a high of 26.1 per 100,000 in 1990 to a low of 17.8 in 2009.

There were also significant variations in the firearm-related death rate for males and females depending on age. For example, in 2010 the male rate was 1.1 at ages five to 14 years, 25 at ages 15 to 24 years, 26.4 at ages 25 to 34 years, and around 20 for all males between the ages of 35 and 74 years. (See Table 6.5.) However, males aged 75 years and older had a higher firearm-related death rate than males of any other age group.

This dramatic increase in the firearm-related death rate among elderly males is in large part a consequence of this group's high rate of suicide. As Table 6.6 shows, at all points between 1950 and 2010 the suicide rate for males aged 65 years and older jumped considerably. In 2010, as in prior years, males were increasingly likely to commit suicide once they reached age 75. Males aged 65 to 74 years had a lower suicide rate (23.9 per 100,000) than middle-aged males, but males aged 75 to 84 years had a higher suicide rate (32.3) than all other younger males, and males aged 85 years and older had the highest suicide rate (47.3) of any demographic subgroup.

As Table 6.5 shows, the firearm-related death rate for females has varied less over time, and in all years the patterns among females by age differed in important ways from the patterns among males. As with males, the firearm-related death rate for females typically increases with age at the transition from childhood to adolescence. Females aged 15 to 24 years had a firearm-related death rate of 2.9 per 100,000 in 2010, which was higher than 0.3 for females aged one to 14 years but much lower than the rate for adolescent males. As with males, the rate for females remained elevated through the middle years of adulthood, at 3.8 for females aged 25 to 44 years and 3.7 for females aged 45 to 64 years. However, in sharp contrast to the pattern among males the firearm-related death rate decreased for females at ages

TABLE 6.2

10 leading causes of nonfatal injuries treated in hospital emergency departments, by age group, 2012

Rank	<1	1–4	5–9	10–14	15–24	25–34	35–44	45–54	55–64	65+	Total
						Age groups					
1	Unintentional fall 145,406	Unintentional fall 928,693	Unintentional fall 682,972	Unintentional struck by/against 612,890	Unintentional struck by/against 1,015,214	Unintentional fall 781,424	Unintentional fall 723,904	Unintentional fall 925,945	Unintentional fall 881,178	Unintentional fall 2,422,463	Unintentional fall 8,974,762
2	Unintentional struck by/against 33,070	Unintentional struck by/against 385,146	Unintentional struck by/against 427,701	Unintentional fall 600,403	Unintentional fall 882,214	Unintentional struck by/against 648,632	Unintentional overexertion 552,464	Unintentional overexertion 480,093	Unintentional struck by/against 269,407	Unintentional struck by/against 280,326	Unintentional struck by/against 4,553,417
3	Unintentional other bite/sting 14,293	Unintentional other bite/sting 172,438	Unintentional cut/pierce 119,092	Unintentional overexertion 318,459	Unintentional overexertion 728,819	Unintentional overexertion 643,571	Unintentional struck by/against 467,962	Unintentional struck by/against 412,908	Unintentional overexertion 259,092	Unintentional overexertion 213,553	Unintentional overexertion 3,385,128
4	Unintentional foreign body 11,245	Unintentional foreign body 145,117	Unintentional other bite/sting 111,305	Unintentional cut/pierce 126,902	Unintentional MV-occupant 666,833	Unintentional MV-occupant 539,425	Unintentional MV-occupant 392,071	Unintentional MV-occupant 358,304	Unintentional MV-occupant 232,972	Unintentional MV-occupant 197,951	Unintentional MV-occupant 2,564,003
5	Unintentional fire/burn 10,837	Unintentional cut/pierce 90,907	Unintentional overexertion 93,654	Unintentional pedal cyclist 98,448	Unintentional cut/pierce 444,644	Unintentional cut/pierce 415,539	Unintentional cut/pierce 310,662	Unintentional other specified 318,190	Unintentional cut/pierce 186,182	Unintentional cut/pierce 152,185	Unintentional cut/pierce 2,145,927
6	Unintentional inhalation/suffocation 9,658	Unintentional overexertion 89,007	Unintentional pedal cyclist 82,935	Unintentional unknown/unspecified 88,748	Other assault* struck by/again 423,804	Other assault* struck by/again 346,347	Unintentional other specified 264,926	Unintentional cut/pierce 293,852	Unintentional other specified 175,130	Unintentional poisoning 99,618	Unintentional other specified 1,580,574
7	Unintentional other specified 9,120	Unintentional other specified 68,237	Unintentional foreign body 64,755	Unintentional MV-occupant 77,363	Unintentional other specified 320,844	Unintentional other specified 294,541	Other assault* struck by/again 217,972	Unintentional poisoning 216,942	Unintentional poisoning 128,260	Unintentional other bite/sting 90,300	Other assault* struck by/again 1,361,096
8	Unintentional overexertion 6,416	Unintentional fire/burn 56,711	Unintentional MV-occupant 60,056	Other assault* struck by/again 71,276	Unintentional other bite/sting 191,062	Unintentional other bite/sting 193,519	Unintentional poisoning 162,002	Other assault* struck by/again 174,556	Unintentional other bite/sting 103,825	Unintentional other specified 79,395	Unintentional other bite/sting 1,250,916
9	Unintentional cut/pierce 5,963	Unintentional unknown/unspecified 45,276	Unintentional dog bite 46,565	Unintentional other bite/sting 69,914	Unintentional unknown/unspecified 154,866	Unintentional poisoning 156,817	Unintentional other bite/sting 148,512	Unintentional other bite/sting 155,748	Other assault* struck by/again 71,819	Unintentional other transport 68,167	Unintentional poisoning 972,923
10	Unintentional unknown/unspecified 5,020	Unintentional dog bite 38,556	Unintentional unknown/unspecified 38,208	Unintentional other transport 49,358	Unintentional poisoning 147,636	Unintentional unknown/unspecified 113,467	Unintentional unknown/unspecified 90,213	Unintentional unknown/unspecified 85,383	Unintentional other transport 52,056	Unintentional unknown/unspecified 62,327	Unintentional unknown/unspecified 734,164

*Includes all assaults that are not classified as sexual assault. It represents the majority of assaults.

MV = Motor vehicle

SOURCE: Adapted from "National Estimates of the 10 Leading Causes of Nonfatal Injuries Treated in Hospital Emergency Departments, United States—2012," in *Ten Leading Causes of Death and Injury*, Centers for Disease Control, 2014, http://www.cdc.gov/injury/wisqars/pdf/leading_cause_of_nonfatal_injury_2012-a.pdf (accessed August 15, 2014)

TABLE 6.3

Nonfatal firearm injuries among children aged 0 to 19 years and rates per 100,000, 2012

Age group	Sex	Number of injuries	Population	Crude rate
00–04	Both sexes	79*	19,999,344	0.39
	Males	55*	10,216,135	0.54
	Females	24*	9,783,209	0.24
05–09	Both sexes	106*	20,475,536	0.52
	Males	0*	10,459,193	0.00
	Females	106*	10,016,343	1.06
10–14	Both sexes	470*	20,669,218	2.28
	Males	332*	10,567,214	3.14
	Females	139*	10,102,004	1.37
15–19	Both sexes	9,399	21,360,702	44.00
	Males	8,493	10,962,861	77.47
	Females	906*	10,397,841	8.71

*Injury estimate is unstable because of small sample size. Use with caution.

SOURCE: Adapted from "Nonfatal Injury Reports, 2001–2012," in *Web-Based Injury Statistics Query and Reporting System (WISQARS)*, Centers for Disease Control and Prevention, 2014, http://webappa.cdc.gov/sasweb/ncipc/nfirates2001.html (accessed July 16, 2014)

TABLE 6.4

BB/pellet gun injuries among children aged 0 to 19 years and rates per 100,000, 2012

Age group	Sex	Number of injuries	Population	Crude rate
00–04	Both sexes	0	19,999,344	0.00
	Males	0	10,216,135	0.00
	Females	0	9,783,209	0.00
05–09	Both sexes	27	20,475,536	0.13
	Males	27	10,459,193	0.26
	Females	0	10,016,343	0.00
10–14	Both sexes	179	20,669,218	0.86
	Males	134	10,567,214	1.27
	Females	45	10,102,004	0.44
15–19	Both sexes	842	21,360,702	3.94
	Males	732	10,962,861	6.68
	Females	110	10,397,841	1.06

SOURCE: Adapted from "Nonfatal Injury Reports, 2001–2012," in *Web-Based Injury Statistics Query and Reporting System (WISQARS)*, Centers for Disease Control and Prevention, 2014, http://webappa.cdc.gov/sasweb/ncipc/nfirates2001.html (accessed July 16, 2014)

75 years and beyond. Females aged 85 years and older had the lowest rate of any adult females, at 1.5.

Part of the reason for the disparity between the firearm-related death rate among elderly males and females is the difference in the two groups' suicide rates. Unlike their elderly male counterparts, women over the age of 75 years have one of the lowest suicide rates of all female age groups. The suicide rate for females aged 85 years and older, 3.3 per 100,000, was nearly three times lower than the highest rate for females aged 45 to 54 years, 9 per 100,000. (See Table 6.6.)

As Table 6.5 shows, African American males at all points between 1970 and 2010 had the highest firearm-related

FIGURE 6.1

Firearm homicides, 1993–2011

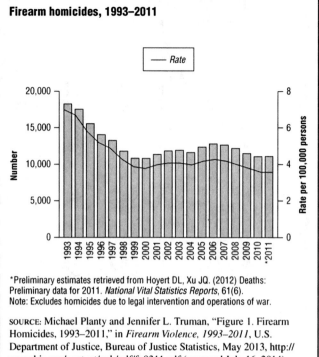

*Preliminary estimates retrieved from Hoyert DL, Xu JQ. (2012) Deaths: Preliminary data for 2011. *National Vital Statistics Reports*, 61(6).
Note: Excludes homicides due to legal intervention and operations of war.

SOURCE: Michael Planty and Jennifer L. Truman, "Figure 1. Firearm Homicides, 1993–2011," in *Firearm Violence, 1993–2011*, U.S. Department of Justice, Bureau of Justice Statistics, May 2013, http://www.bjs.gov/content/pub/pdf/fv9311.pdf (accessed July 16, 2014)

death rate among all male demographic subgroups. The age-adjusted firearm-related death rate for African American males fell steadily between 1970 and 2010, from 70.8 per 100,000 to 31.8 per 100,000. Throughout this period, however, African American males typically died from firearm-related injuries at rates two to three times higher than Hispanic males, white males, and Native American or Alaskan Native males, and at rates five to seven times higher than Asian or Pacific Islander males, who had the lowest firearm-related death rate of any group. Although African American men aged 25 to 34 years had the highest firearm-related death rate of any demographic subgroup in 1970 (at 145.6 per 100,000), as the firearm homicide rate for all groups peaked during the mid-1990s, African American males aged 15 to 24 years had the highest rate of death from firearm-related injuries, at 138 in 1990 and 138.7 in 1995. In 2010 African American males between the ages of 15 and 24 years had a firearm-related death rate of 73.2, and African American males aged 25 to 34 years had a rate of 78.2. In other words, African American males aged 15 to 34 years were over four times more likely to be killed by firearms than the average American male, whose age-adjusted rate in 2010 was 17.9.

Table 6.7 shows firearm-related deaths and rates attributable to homicide and suicide between 2002 and 2011, and Table 6.8 shows the firearm-related deaths and rates attributable to unintentional shootings between 2002 and 2011. As is evident, intentional firearm-related deaths are much more common than unintentional ones.

TABLE 6.5

Death rates for firearm-related injuries, by sex, race, Hispanic origin, and age, selected years 1970–2010

[Data are based on death certificates]

Sex, race, Hispanic origin, and age	1970[a]	1980[a]	1990[a]	1995[a]	2000[b]	2005[b]	2009[b]	2010[b]
All persons				Deaths per 100,000 resident population				
All ages, age-adjusted[c]	14.3	14.8	14.6	13.4	10.2	10.3	10.1	10.1
All ages, crude	13.1	14.9	14.9	13.5	10.2	10.4	10.2	10.3
Under 1 year	*	*	*	*	*	*	*	*
1–14 years	1.6	1.4	1.5	1.6	0.7	0.7	0.6	0.6
1–4 years	1.0	0.7	0.6	0.6	0.3	0.4	0.4	0.4
5–14 years	1.7	1.6	1.9	1.9	0.9	0.8	0.7	0.7
15–24 years	15.5	20.6	25.8	26.7	16.8	16.1	14.4	14.2
15–19 years	11.4	14.7	23.3	24.1	12.9	12.2	11.1	10.6
20–24 years	20.3	26.4	28.1	29.2	20.9	20.0	18.0	17.9
25–44 years	20.9	22.5	19.3	16.9	13.1	13.8	13.2	13.3
25–34 years	22.2	24.3	21.8	19.6	14.5	16.1	14.5	15.0
35–44 years	19.6	20.0	16.3	14.3	11.9	11.7	11.9	11.7
45–64 years	17.6	15.2	13.6	11.7	10.0	10.6	11.4	11.6
45–54 years	18.1	16.4	13.9	12.0	10.5	11.2	11.8	12.0
55–64 years	17.0	13.9	13.3	11.3	9.4	9.7	10.8	11.1
65 years and over	13.8	13.5	16.0	14.1	12.2	11.8	11.9	11.7
65–74 years	14.5	13.8	14.4	12.8	10.6	10.2	10.9	10.7
75–84 years	13.4	13.4	19.4	16.3	13.9	13.6	13.3	12.7
85 years and over	10.2	11.6	14.7	14.4	14.2	13.0	12.5	13.2
Male								
All ages, age-adjusted[c]	24.8	25.9	26.1	23.8	18.1	18.5	17.8	17.9
All ages, crude	22.2	25.7	26.2	23.6	17.8	18.4	17.9	18.0
Under 1 year	*	*	*	*	*	*	*	*
1–14 years	2.3	2.0	2.2	2.3	1.1	1.0	0.9	1.0
1–4 years	1.2	0.9	0.7	0.8	0.4	0.5	0.5	0.6
5–14 years	2.7	2.5	2.9	2.9	1.4	1.2	1.0	1.1
15–24 years	26.4	34.8	44.7	46.5	29.4	28.5	25.3	25.0
15–19 years	19.2	24.5	40.1	41.6	22.4	21.5	19.3	18.4
20–24 years	35.1	45.2	49.1	51.5	37.0	35.7	31.6	31.8
25–44 years	34.1	38.1	32.6	28.4	22.0	23.7	22.4	22.9
25–34 years	36.5	41.4	37.0	33.2	24.9	28.2	25.0	26.4
35–44 years	31.6	33.2	27.4	23.6	19.4	19.5	19.9	19.3
45–64 years	31.0	25.9	23.4	20.0	17.1	18.2	19.3	19.9
45–54 years	30.7	27.3	23.2	20.1	17.6	18.9	19.6	20.3
55–64 years	31.3	24.5	23.7	19.8	16.3	17.2	19.1	19.3
65 years and over	29.7	29.7	35.3	30.7	26.4	25.1	24.8	24.1
65–74 years	29.5	27.8	28.2	25.1	20.3	19.3	20.6	20.0
75–84 years	31.0	33.0	46.9	37.8	32.2	30.5	28.8	27.5
85 years and over	26.2	34.9	49.3	47.1	44.7	39.3	35.7	37.4
Female								
All ages, age-adjusted[c]	4.8	4.7	4.2	3.8	2.8	2.7	2.8	2.7
All ages, crude	4.4	4.7	4.3	3.8	2.8	2.7	2.8	2.7
Under 1 year	*	*	*	*	*	*	*	*
1–14 years	0.8	0.7	0.8	0.8	0.3	0.4	0.3	0.3
1–4 years	0.9	0.5	0.5	0.5	*	0.3	0.4	0.3
5–14 years	0.8	0.7	1.0	0.9	0.4	0.4	0.3	0.3
15–24 years	4.8	6.1	6.0	5.9	3.5	3.0	3.1	2.9
15–19 years	3.5	4.6	5.7	5.6	2.9	2.4	2.4	2.3
20–24 years	6.4	7.7	6.3	6.1	4.2	3.6	3.7	3.5
25–44 years	8.3	7.4	6.1	5.5	4.2	3.9	3.9	3.8
25–34 years	8.4	7.5	6.7	5.8	4.0	3.8	3.9	3.5
35–44 years	8.2	7.2	5.4	5.2	4.4	4.0	3.9	4.1
45–64 years	5.4	5.4	4.5	3.9	3.4	3.3	3.8	3.7
45–54 years	6.4	6.2	4.9	4.2	3.6	3.7	4.3	3.8
55–64 years	4.2	4.6	4.0	3.5	3.0	2.8	3.2	3.4
65 years and over	2.4	2.5	3.1	2.8	2.2	2.1	2.2	2.2
65–74 years	2.8	3.1	3.6	3.0	2.5	2.5	2.6	2.6
75–84 years	1.7	1.7	2.9	2.8	2.0	2.1	2.2	2.1
85 years and over	*	1.3	1.3	1.8	1.7	1.4	1.3	1.5

In terms of the age-adjusted rates of death, both intentional and unintentional firearm deaths saw modest overall declines between 2002 and 2011. The raw number of firearm-related homicides and suicides rose slightly during this period, but the growth of the U.S. population outpaced this increase, so the rate declined slightly. (See Table 6.7.)

Firearm Suicides versus Homicides

In "Suicides Account for Most Gun Deaths" (May 24, 2013, http://www.pewresearch.org/fact-tank/2013/05/24/suicides-account-for-most-gun-deaths), Drew DeSilver of the Pew Research Center notes that "since the CDC began publishing data in 1981, gun suicides have

TABLE 6.5

Death rates for firearm-related injuries, by sex, race, Hispanic origin, and age, selected years 1970–2010 [CONTINUED]

[Data are based on death certificates]

Sex, race, Hispanic origin, and age	1970[a]	1980[a]	1990[a]	1995[a]	2000[b]	2005[b]	2009[b]	2010[b]
				Deaths per 100,000 resident population				
White male[d]								
All ages, age-adjusted[c]	19.7	22.1	22.0	20.1	15.9	15.9	15.9	16.1
All ages, crude	17.6	21.8	21.8	19.9	15.6	16.0	16.2	16.5
1–14 years	1.8	1.9	1.9	1.9	1.0	0.8	0.8	0.8
15–24 years	16.9	28.4	29.5	30.8	19.6	18.3	16.9	16.2
25–44 years	24.2	29.5	25.7	23.2	18.0	18.4	18.1	18.6
25–34 years	24.3	31.1	27.8	25.2	18.1	19.4	17.9	19.1
35–44 years	24.1	27.1	23.3	21.2	17.9	17.5	18.2	18.0
45–64 years	27.4	23.3	22.8	19.5	17.4	19.0	20.5	21.3
65 years and over	29.9	30.1	36.8	32.2	28.2	27.0	26.9	26.5
Black or African American male[d]								
All ages, age-adjusted[c]	70.8	60.1	56.3	49.2	34.2	36.7	32.3	31.8
All ages, crude	60.8	57.7	61.9	52.9	36.1	38.6	33.7	33.4
1–14 years	5.3	3.0	4.4	4.4	1.8	2.1	1.6	1.9
15–24 years	97.3	77.9	138.0	138.7	89.3	86.2	72.8	73.2
25–44 years	126.2	114.1	90.3	70.2	54.1	64.8	57.2	57.3
25–34 years	145.6	128.4	108.6	92.3	74.8	92.1	76.0	78.2
35–44 years	104.2	92.3	66.1	46.3	34.3	38.6	37.7	35.2
45–64 years	71.1	55.6	34.5	28.3	18.4	17.2	17.1	16.5
65 years and over	30.6	29.7	23.9	21.8	13.8	13.5	12.1	9.4
American Indian or Alaska Native male[d]								
All ages, age-adjusted[c]	—	24.0	19.4	19.4	13.1	14.4	11.4	11.7
All ages, crude	—	27.5	20.5	20.9	13.2	14.9	11.6	12.5
15–24 years	—	55.3	49.1	40.9	26.9	28.6	22.2	26.0
25–44 years	—	43.9	25.4	31.2	16.6	21.3	16.3	16.9
45–64 years	—	*	*	14.2	12.2	12.1	11.1	11.1
65 years and over	—	*	*	*	*	*	*	*
Asian or Pacific Islander male[d]								
All ages, age-adjusted[c]	—	7.8	8.8	9.2	6.0	5.1	4.4	4.2
All ages, crude	—	8.2	9.4	10.0	6.2	5.4	4.5	4.4
15–24 years	—	10.8	21.0	24.3	9.3	10.9	5.4	6.8
25–44 years	—	12.8	10.9	10.6	8.1	6.4	6.2	6.0
45–64 years	—	10.4	8.1	8.2	7.4	5.7	5.2	4.4
65 years and over	—	*	*	*	*	*	4.7	3.9
Hispanic or Latino male[d, e]								
All ages, age-adjusted[c]	—	—	27.6	23.8	13.6	13.4	11.4	10.5
All ages, crude	—	—	29.9	26.2	14.2	14.3	11.4	10.5
1–14 years	—	—	2.6	2.8	1.0	0.7	0.7	0.6
15–24 years	—	—	55.5	61.7	30.8	30.8	23.3	20.9
25–44 years	—	—	42.7	31.4	17.3	19.7	15.5	14.4
25–34 years	—	—	47.3	36.4	20.3	24.4	18.0	18.0
35–44 years	—	—	35.4	24.2	13.2	13.9	12.4	10.2
45–64 years	—	—	21.4	17.2	12.0	9.2	9.5	9.1
65 years and over	—	—	19.1	16.5	12.2	10.2	10.8	9.9
White, not Hispanic or Latino male[e]								
All ages, age-adjusted[c]	—	—	20.6	18.6	15.5	15.5	16.1	16.6
All ages, crude	—	—	20.4	18.5	15.7	16.1	17.0	17.6
1–14 years	—	—	1.6	1.6	1.0	0.9	0.7	0.9
15–24 years	—	—	24.1	23.5	16.2	14.1	14.2	14.2
25–44 years	—	—	23.3	21.4	17.9	17.7	18.4	19.4
25–34 years	—	—	24.7	22.5	17.2	17.4	17.4	18.9
35–44 years	—	—	21.6	20.4	18.4	17.9	19.4	19.9
45–64 years	—	—	22.7	19.5	17.8	20.0	21.8	22.8
65 years and over	—	—	37.4	32.5	29.0	28.1	27.9	27.6
White female[d]								
All ages, age-adjusted[c]	4.0	4.2	3.8	3.5	2.7	2.6	2.8	2.7
All ages, crude	3.7	4.1	3.8	3.5	2.7	2.6	2.9	2.8
15–24 years	3.4	5.1	4.8	4.5	2.8	2.3	2.4	2.3
25–44 years	6.9	6.2	5.3	4.9	3.9	3.8	3.8	3.7
45–64 years	5.0	5.1	4.5	4.0	3.5	3.5	4.2	4.1
65 years and over	2.2	2.5	3.1	2.8	2.4	2.3	2.5	2.5

TABLE 6.5

Death rates for firearm-related injuries, by sex, race, Hispanic origin, and age, selected years 1970–2010 [CONTINUED]

[Data are based on death certificates]

Sex, race, Hispanic origin, and age	1970[a]	1980[a]	1990[a]	1995[a]	2000[b]	2005[b]	2009[b]	2010[b]
Black or African American female[d]			Deaths per 100,000 resident population					
All ages, age-adjusted[c]	11.1	8.7	7.3	6.2	3.9	3.6	3.4	3.3
All ages, crude	10.0	8.8	7.8	6.5	4.0	3.7	3.5	3.3
15–24 years	15.2	12.3	13.3	13.2	7.6	6.6	6.7	6.4
25–44 years	19.4	16.1	12.4	9.8	6.5	6.0	5.7	5.6
45–64 years	10.2	8.2	4.8	4.1	3.1	2.7	2.4	2.2
65 years and over	4.3	3.1	3.1	2.6	1.3	1.3	*	*
American Indian or Alaska Native female[d]								
All ages, age-adjusted[c]	—	5.8	3.3	3.8	2.9	2.2	2.5	2.6
All ages, crude	—	5.8	3.4	4.1	2.9	2.3	2.4	2.4
15–24 years	—	*	*	*	*	*	*	*
25–44 years	—	10.2	*	7.0	5.5	*	3.6	3.7
45–64 years	—	*	*	*	*	*	*	*
65 years and over	—	*	*	*	*	*	*	*
Asian or Pacific Islander female[d]								
All ages, age-adjusted[c]	—	2.0	1.9	2.0	1.1	0.9	0.9	0.6
All ages, crude	—	2.1	2.1	2.1	1.2	0.9	0.9	0.6
15–24 years	—	*	*	3.9	*	2.0	*	*
25–44 years	—	3.2	2.7	2.7	1.5	0.9	1.1	1.1
45–64 years	—	*	*	*	*	*	1.2	*
65 years and over	—	*	*	*	*	*	*	*
Hispanic or Latina female[d, e]								
All ages, age-adjusted[c]	—	—	3.3	3.1	1.8	1.5	1.4	1.3
All ages, crude	—	—	3.6	3.3	1.8	1.5	1.4	1.3
15–24 years	—	—	6.9	6.1	2.9	2.4	2.6	2.1
25–44 years	—	—	5.1	4.7	2.5	2.6	1.9	1.8
45–64 years	—	—	2.4	2.4	2.2	1.2	1.4	1.5
65 years and over	—	—	*	*	*	*	*	*
White, not Hispanic or Latina female[e]								
All ages, age-adjusted[c]	—	—	3.7	3.4	2.8	2.7	3.0	3.0
All ages, crude	—	—	3.7	3.5	2.9	2.8	3.1	3.1
15–24 years	—	—	4.3	4.1	2.7	2.2	2.2	2.3
25–44 years	—	—	5.1	4.8	4.2	4.0	4.3	4.2
45–64 years	—	—	4.6	4.1	3.6	3.8	4.5	4.4
65 years and over	—	—	3.2	2.8	2.4	2.4	2.6	2.6

*Rates based on fewer than 20 deaths are considered unreliable and are not shown.

—Data not available.

[a]Underlying cause of death was coded according to the 8th Revision of the International Classification of Diseases (ICD) in 1970 and 9th Revision in 1980–1998.

[b]Starting with 1999 data, cause of death is coded according to ICD-10.

[c]Age-adjusted rates are calculated using the year 2000 standard population. Prior to 2001, age-adjusted rates were calculated using standard million proportions based on rounded population numbers. Starting with 2001 data, unrounded population numbers are used to calculate age-adjusted rates.

[d]The race groups, white, black, Asian or Pacific Islander, and American Indian or Alaska Native, include persons of Hispanic and non-Hispanic origin. Persons of Hispanic origin may be of any race. Death rates for Hispanic, American Indian and Alaska Native, and Asian or Pacific Islander persons should be interpreted with caution because of inconsistencies in reporting Hispanic origin or race on the death certificate (death rate numerators) compared with population figures (death rate denominators). The net effect of misclassification is an underestimation of deaths and death rates for races other than white and black.

[e]Prior to 1997, data from states that did not report Hispanic origin on the death certificate were excluded.

Notes: Starting with *Health, United States, 2003*, rates for 1991–1999 were revised using intercensal population estimates based on the 1990 and 2000 censuses. For 2000, population estimates are bridged-race April 1 census counts. Starting with *Health, United States, 2012*, rates for 2001–2009 were revised using intercensal population estimates based on the 2000 and 2010 censuses. For 2010, population estimates are bridged-race April 1 census counts. Age groups were selected to minimize the presentation of unstable age-specific death rates based on small numbers of deaths and for consistency among comparison groups.

Starting with 2003 data, some states allowed the reporting of more than one race on the death certificate. The multiple-race data for these states were bridged to the single-race categories of the 1977 Office of Management and Budget standards, for comparability with other states.

SOURCE: "Table 36. Death Rates for Firearm-Related Injuries, by Sex, Race, Hispanic Origin, and Age: United States, Selected Years 1970–2010," in *Health, United States, 2013: With Special Feature on Prescription Drugs*, Centers for Disease Control and Prevention, National Center for Health Statistics, May 2014, http://www.cdc.gov/nchs/data/hus/hus13.pdf (accessed July 16, 2014)

outnumbered gun homicides. But as gun homicides have declined sharply in recent years, suicides have become a greater share of all firearm deaths." By 2010 firearm suicides accounted for 61% of all gun deaths and firearm homicides accounted for 35% of all gun deaths. The rate at which people committed suicide with firearms was 6.3 per 100,000 in 2010, and the rate at which people committed homicide with firearms was 3.6 per 100,000. (See Figure 6.2.) According to DeSilver, males account for 87% of firearm suicides, and people aged 65 years and older have the highest suicide rate among all age groups, at 10.6 per 100,000.

TABLE 6.6

Death rates for suicide, by sex and age, selected years 1950–2010

[Data are based on death certificates]

Sex, race, Hispanic origin, and age	1950[a]	1960[a]	1970	1980	1990	2000[b]	2009[b]	2010[b]
All persons				Deaths per 100,000 resident population				
All ages, age-adjusted[c]	13.2	12.5	13.1	12.2	12.5	10.4	11.8	12.1
All ages, crude	11.4	10.6	11.6	11.9	12.4	10.4	12.0	12.4
Under 1 year	—	—	—	—	—	—	—	—
1–4 years	—	—	—	—	—	—	—	—
5–14 years	0.2	0.3	0.3	0.4	0.8	0.7	0.6	0.7
15–24 years	4.5	5.2	8.8	12.3	13.2	10.2	10.0	10.5
15–19 years	2.7	3.6	5.9	8.5	11.1	8.0	7.5	7.5
20–24 years	6.2	7.1	12.2	16.1	15.1	12.5	12.6	13.6
25–44 years	11.6	12.2	15.4	15.6	15.2	13.4	14.6	15.0
25–34 years	9.1	10.0	14.1	16.0	15.2	12.0	13.1	14.0
35–44 years	14.3	14.2	16.9	15.4	15.3	14.5	16.1	16.0
45–64 years	23.5	22.0	20.6	15.9	15.3	13.5	17.9	18.6
45–54 years	20.9	20.7	20.0	15.9	14.8	14.4	19.2	19.6
55–64 years	26.8	23.7	21.4	15.9	16.0	12.1	16.4	17.5
65 years and over	30.0	24.5	20.8	17.6	20.5	15.2	14.8	14.9
65–74 years	29.6	23.0	20.8	16.9	17.9	12.5	13.7	13.7
75–84 years	31.1	27.9	21.2	19.1	24.9	17.6	15.8	15.7
85 years and over	28.8	26.0	19.0	19.2	22.2	19.6	16.4	17.6
Male								
All ages, age-adjusted[c]	21.2	20.0	19.8	19.9	21.5	17.7	19.2	19.8
All ages, crude	17.8	16.5	16.8	18.6	20.4	17.1	19.3	19.9
Under 1 year	—	—	—	—	—	—	—	—
1–4 years	—	—	—	—	—	—	—	—
5–14 years	0.3	0.4	0.5	0.6	1.1	1.2	0.8	0.9
15–24 years	6.5	8.2	13.5	20.2	22.0	17.1	16.1	16.9
15–19 years	3.5	5.6	8.8	13.8	18.1	13.0	11.6	11.7
20–24 years	9.3	11.5	19.3	26.8	25.7	21.4	20.8	22.2
25–44 years	17.2	17.9	20.9	24.0	24.4	21.3	23.0	23.6
25–34 years	13.4	14.7	19.8	25.0	24.8	19.6	21.0	22.5
35–44 years	21.3	21.0	22.1	22.5	23.9	22.8	24.9	24.6
45–64 years	37.1	34.4	30.0	23.7	24.3	21.3	27.9	29.2
45–54 years	32.0	31.6	27.9	22.9	23.2	22.4	29.3	30.4
55–64 years	43.6	38.1	32.7	24.5	25.7	19.4	26.1	27.7
65 years and over	52.8	44.0	38.4	35.0	41.6	31.1	29.1	29.0
65–74 years	50.5	39.6	36.0	30.4	32.2	22.7	24.3	23.9
75–84 years	58.3	52.5	42.8	42.3	56.1	38.6	32.9	32.3
85 years and over	58.3	57.4	42.4	50.6	65.9	57.5	44.0	47.3
Female								
All ages, age-adjusted[c]	5.6	5.6	7.4	5.7	4.8	4.0	4.9	5.0
All ages, crude	5.1	4.9	6.6	5.5	4.8	4.0	5.0	5.2
Under 1 year	—	—	—	—	—	—	—	—
1–4 years	—	—	—	—	—	—	—	—
5–14 years	0.1	0.1	0.2	0.2	0.4	0.3	0.5	0.4
15–24 years	2.6	2.2	4.2	4.3	3.9	3.0	3.6	3.9
15–19 years	1.8	1.6	2.9	3.0	3.7	2.7	3.2	3.1
20–24 years	3.3	2.9	5.7	5.5	4.1	3.2	4.1	4.7
25–44 years	6.2	6.6	10.2	7.7	6.2	5.4	6.2	6.4
25–34 years	4.9	5.5	8.6	7.1	5.6	4.3	5.1	5.3
35–44 years	7.5	7.7	11.9	8.5	6.8	6.4	7.4	7.5
45–64 years	9.9	10.2	12.0	8.9	7.1	6.2	8.5	8.6
45–54 years	9.9	10.2	12.6	9.4	6.9	6.7	9.3	9.0
55–64 years	9.9	10.2	11.4	8.4	7.3	5.4	7.4	8.0
65 years and over	9.4	8.4	8.1	6.1	6.4	4.0	4.0	4.2
65–74 years	10.1	8.4	9.0	6.5	6.7	4.0	4.6	4.8
75–84 years	8.1	8.9	7.0	5.5	6.3	4.0	3.6	3.7
85 years and over	8.2	6.0	5.9	5.5	5.4	4.2	3.2	3.3

As Table 6.9 shows, firearm homicides and firearm suicides were among the 10 leading causes of injury deaths (deaths caused by some external force, as opposed to those caused by diseases or congenital ailments) for almost all age groups in the United States in 2011. Firearm homicides were among the top-10 causes of injury deaths for every age group except those under age one and those aged 65 years and older, and firearm suicides were among the top-10 causes of injury deaths for every age group except those under age one, between one and four years, and between five and nine years. Among those aged 10 to 14 years, 15 to 24 years, and 25 to 34 years, more people died of firearm homicides than firearm suicides. Among those aged 35 to 44 years, 45 to 54 years, 55 to 64 years, and 65 years and older, far more died of firearm suicides than firearm homicides. In all,

TABLE 6.6

Death rates for suicide, by sex and age, selected years 1950–2010 [CONTINUED]

[Data are based on death certificates]

—Category not applicable.
[a]Includes deaths of persons who were not residents of the 50 states and the District of Columbia (D.C.).
[b]Starting with 1999 data, cause of death is coded according to ICD–10.
[c]Age-adjusted rates are calculated using the year 2000 standard population. Prior to 2001, age-adjusted rates were calculated using standard million proportions based on rounded population numbers. Starting with 2001 data, unrounded population numbers are used to calculate age-adjusted rates.
Notes: Starting with *Health, United States, 2003*, rates for 1991–1999 were revised using intercensal population estimates based on the 1990 and 2000 censuses. For 2000, population estimates are bridged-race April 1 census counts. Starting with *Health, United States, 2012*, rates for 2001–2009 were revised using intercensal population estimates based on the 2000 and 2010 censuses. For 2010, population estimates are bridged-race April 1 census counts. Figures for 2001 include September 11-related deaths for which death certificates were filed as of October 24, 2002. Age groups were selected to minimize the presentation of unstable age-specific death rates based on small numbers of deaths and for consistency among comparison groups. Starting with 2003 data, some states allowed the reporting of more than one race on the death certificate. The multiple-race data for these states were bridged to the single-race categories of the 1977 Office of Management and Budget standards, for comparability with other states.

SOURCE: Adapted from "Table 35. Death Rates for Suicide, by Sex, Race, Hispanic Origin, and Age: United States, Selected Years 1950–2010," in *Health, United States, 2013: With Special Feature on Prescription Drugs*, Centers for Disease Control and Prevention, National Center for Health Statistics, May 2014, http://www.cdc.gov/nchs/data/hus/hus13.pdf (accessed July 16, 2014)

TABLE 6.7

Fatal gunshot (homicide and suicide) deaths and rates per 100,000 people, 2002–11

Year	Deaths	Population	Age-adjusted rate
2002	29,237	287,625,193	10.09
2003	29,174	290,107,933	9.96
2004	28,685	292,805,298	9.69
2005	29,684	295,516,599	9.93
2006	30,034	298,379,912	9.94
2007	30,335	301,231,207	9.95
2008	30,728	304,093,966	9.95
2009	30,561	306,771,529	9.80
2010	30,814	308,745,538	9.79
2011	31,512	311,587,816	9.90
Total	**300,764**	**2,996,864,991**	—

SOURCE: Adapted from "Fatal Injury Reports, National and Regional, 1999–2011," in *Web-Based Injury Statistics Query and Reporting System (WISQARS)*, Centers for Disease Control and Prevention, 2014, http://webappa.cdc.gov/sasweb/ncipc/mortrate10_us.html (accessed July 16, 2014)

TABLE 6.8

Unintentional firearm deaths and rates per 100,000 people, 2002–11

Year	Deaths	Population	Age-adjusted rate
2002	762	287,625,193	0.26
2003	730	290,107,933	0.25
2004	649	292,805,298	0.22
2005	789	295,516,599	0.27
2006	642	298,379,912	0.21
2007	613	301,231,207	0.20
2008	592	304,093,966	0.19
2009	554	306,771,529	0.18
2010	606	308,745,538	0.20
2011	591	311,587,816	0.19
Total	**6,528**	**2,996,864,991**	—

SOURCE: Adapted from "Fatal Injury Reports, National and Regional, 1999–2011," in *Web-Based Injury Statistics Query and Reporting System (WISQARS)*, Centers for Disease Control and Prevention, 2014, http://webappa.cdc.gov/sasweb/ncipc/mortrate10_us.html (accessed July 16, 2014)

firearm suicides accounted for 19,990 deaths in 2011, and firearm homicides accounted for 11,068 deaths.

Michael L. Nance et al. examine in "Variation in Pediatric and Adolescent Firearm Mortality Rates in Rural and Urban US Counties" (*Pediatrics*, vol. 125, no. 6, June 2010) whether children and youth died from firearms at varying rates depending on whether they lived in rural, suburban, or urban settings. The researchers report that their analysis of death data files from the National Center for Health Statistics' National Vital Statistics System between 1999 and 2006 reveal that children and youth aged zero to 19 years in the most rural counties had firearm mortality rates that were equivalent to those for children and youth of the same age range in the most urban counties. The type of death varied, however. In urban counties youth firearm deaths were more frequently related to homicide, whereas in rural counties youth firearm deaths were more frequently related to suicide and unintentional gunshot wounds.

THE COST OF FIREARM INJURIES

In "Medical Costs and Productivity Losses Due to Interpersonal and Self-Directed Violence in the United States" (*American Journal of Preventive Medicine*, vol. 32, no. 6, 2007), Phaedra S. Corso et al. state that the estimated total lifetime cost of firearm injuries due to assaults occurring in 2000 was about $17.4 billion—$822 million for medical treatment and $16.6 billion for lost work and household productivity. Costs were much higher for males than for females, with males accounting for 90% of the total costs. The estimated total lifetime cost of firearm injuries due to assaults occurring in 2000 for males was $15.7 billion—$734 million for medical treatment and $14.9 billion for lost productivity—and for females was $1.8 billion—$88 million for medical treatment and $1.7 billion for lost productivity.

FIGURE 6.2

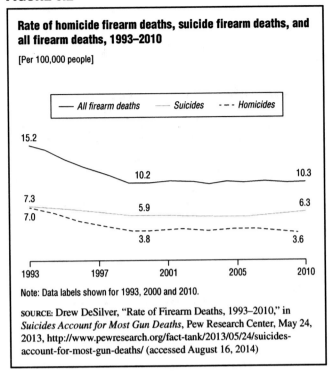

Rate of homicide firearm deaths, suicide firearm deaths, and all firearm deaths, 1993–2010

[Per 100,000 people]

— All firearm deaths   ⋯⋯ Suicides   - - - Homicides

15.2

10.2   10.3

7.3   5.9   6.3

7.0

3.8   3.6

1993   1997   2001   2005   2010

Note: Data labels shown for 1993, 2000 and 2010.

SOURCE: Drew DeSilver, "Rate of Firearm Deaths, 1993–2010," in *Suicides Account for Most Gun Deaths*, Pew Research Center, May 24, 2013, http://www.pewresearch.org/fact-tank/2013/05/24/suicides-account-for-most-gun-deaths/ (accessed August 16, 2014)

Corso et al. also indicate that the estimated total lifetime cost of self-inflicted (suicide) firearm injuries occurring in 2000 was only about $1 billion less, at $16.4 billion—$124 million for medical treatment and $16.3 billion for lost productivity. Once again, costs were much higher for males than for females, with males accounting for 90% of the total costs. The estimated total lifetime cost of self-inflicted firearm injuries occurring in 2000 for males was $14.9 billion—$101 million for medical treatment and $14.8 billion for lost productivity—and for females was $1.6 billion—$23 million for medical treatment and $1.6 billion for lost productivity.

Notice that the proportion of the costs for medical treatment versus lost productivity was lower for the self-inflicted injuries than for the assault injuries. This difference is attributable to the fact that the death rate for self-inflicted firearm injuries is higher than the death rate for firearm injuries due to assaults. Medical treatment costs are low with a high proportion of deaths, but costs due to lost productivity are very high. Corso et al. reveal that in 2000 the overall fatal rate for firearm injuries because of assault was 4 per 100,000 population, whereas the overall fatal rate for self-inflicted firearm injuries was 6 per 100,000 population. For men, the rates were 7 per 100,000 population when firearm injuries were inflicted during assaults and 11 per 100,000 population when self-inflicted. For women, the rates were 1 per 100,000 population and 2 per 100,000 population, respectively.

In *The Hospital Costs of Firearm Assaults* (September 2013, http://www.urban.org/UploadedPDF/412894-

The-Hospital-Costs-of-Firearm-Assaults.pdf), Emily M. Howell and Peter Abraham of the Urban Institute attempt to quantify how much money is spent on emergency department (ED) and inpatient hospital care resulting from both fatal and nonfatal firearm assaults. The researchers note that in 2010 there were 36,341 ED admissions and 25,024 hospitalizations resulting from firearm assault incidents. An average ED visit for firearm assault injuries cost $1,126 in 2010, for a total ED cost of $40.9 million; and an average inpatient hospital visit for firearm assault injuries cost $23,497, for a total inpatient cost of $588 million. The total ED and hospital bills generated by the victims of gun violence in 2010 was thus $628.9 million.

Howell and Abraham note that young males living in the poorest areas of the country account for much of the hospital spending on firearm assault injuries. Males accounted for 91% of hospital costs related to firearm injuries in 2010, with males aged 15 to 24 years accounting for 41% and males aged 25 to 34 years accounting for 28%. People living in the poorest zip codes, those in the lowest quartile (the bottom 25%) by household income, accounted for 51% of hospital costs from firearm assault injuries, and those living in the next lowest quartile accounted for 22% of hospital costs. People receiving health care through government programs such as Medicaid (a state and federal health insurance program for low-income people) and Medicare (a federal health insurance program for people aged 65 years and older and people with disabilities) accounted for more than half (52%) of total hospital costs relating to firearm injuries, and people without any form of insurance coverage accounted for 28% of these costs. Thus, 80% of firearm violence costs in hospitals are passed on to the federal government and ultimately to taxpayers.

GUNS AND SELF-DEFENSE

The frequency with which guns are used in self-defense is one of many topics hotly disputed by both gun control and gun rights advocates. Most research suggests that the phenomenon of firearm use in self-defense is exceptionally rare, but gun rights groups typically question the neutrality of such studies and contend that guns are used in self-defense with relative frequency.

As discussed in Chapter 1, gun rights advocates often point to the research of Gary Kleck, a criminologist at Florida State University, in suggesting that guns prevent crime more often than they are used to commit crime. Kleck has consistently asserted, over the course of a number of articles and books written with various coauthors, that guns are used in self-defense approximately 2 million to 2.5 million times annually. Kleck's estimate suggests that guns are used more often in self-defense than to commit crimes, and if this assertion is accurate, it

TABLE 6.9

10 leading causes of injury deaths, by age group, 2011

Rank	<1	1–4	5–9	10–14	15–24	25–34	35–44	45–54	55–64	65+	Total
					Age groups						
1	Unintentional suffocation 896	Unintentional drowning 438	Unintentional MV traffic 350	Unintentional MV traffic 437	Unintentional MV traffic 6,926	Unintentional poisoning 7,652	Unintentional poisoning 8,075	Unintentional poisoning 10,379	Unintentional poisoning 5,048	Unintentional fall 22,901	Unintentional poisoning 36,280
2	Homicide unspecified 142	Unintentional MV traffic 330	Unintentional drowning 128	Suicide suffocation 177	Homicide firearm 3,825	Unintentional MV traffic 5,569	Unintentional MV traffic 4,425	Unintentional MV traffic 5,240	Unintentional MV traffic 4,184	Unintentional MV traffic 6,225	Unintentional MV traffic 33,783
3	Unintentional MV traffic 93	Homicide unspecified 181	Unintentional fire/burn 81	Homicide firearm 107	Unintentional poisoning 3,440	Homicide firearm 3,271	Suicide firearm 2,837	Suicide firearm 4,100	Suicide firearm 3,522	Unintentional unspecified 4,630	Unintentional fall 27,483
4	Homicide other spec., classifiable 82	Unintentional suffocation 144	Homicide firearm 55	Unintentional drowning 107	Suicide firearm 2,168	Suicide firearm 2,740	Suicide suffocation 1,959	Suicide suffocation 2,062	Unintentional fall 2,141	Suicide firearm 4,526	Suicide firearm 19,990
5	Unintentional drowning 52	Unintentional fire/burn 130	Unintentional suffocation 34	Suicide firearm 91	Suicide suffocation 1,898	Suicide suffocation 2,055	Homicide firearm 1,718	Suicide poisoning 1,946	Suicide poisoning 1,411	Unintentional suffocation 3,402	Homicide firearm 11,068
6	Undetermined suffocation 40	Homicide other spec., classifiable 92	Unintentional other land transport 31	Unintentional other land transport 47	Unintentional drowning 543	Suicide poisoning 816	Suicide poisoning 1,280	Unintentional fall 1,368	Suicide suffocation 1,107	Adverse effects 1,628	Suicide suffocation 9,913
7	Unintentional unspecified 28	Unintentional pedestrian, other 88	Unintentional natural/environment 28	Unintentional suffocation 43	Homicide cut/pierce 395	Undetermined poisoning 594	Undetermined poisoning 655	Homicide firearm 1,147	Unintentional suffocation 644	Unintentional poisoning 1,581	Suicide poisoning 6,564
8	Unintentional fire/burn 24	Unintentional struck by or against 56	Unintentional firearm 16	Unintentional fire/burn 42	Suicide poisoning 349	Homicide cut/pierce 447	Unintentional fall 524	Undetermined poisoning 952	Homicide firearm 546	Unintentional fire/burn 1,073	Unintentional suffocation 6,242
9	Homicide suffocation 23	Homicide firearm 48	Unintentional pedestrian, other 16	Unintentional poisoning 35	Undetermined poisoning 252	Unintentional drowning 442	Unintentional drowning 414	Unintentional suffocation 536	Unintentional fire/burn 522	Unintentional natural/environment 825	Unintentional unspecified 5,871
10	Unintentional natural/environment 21	Unintentional natural/environment 40	Two tied 15	Unintentional other transport 32	Unintentional fall 205	Unintentional fall 279	Homicide cut/pierce 317	Unintentional drowning 479	Unintentional unspecified 496	Suicide poisoning 751	Unintentional drowning 3,556

MV = Motor vehicle

SOURCE: Adapted from "10 Leading Causes of Injury Deaths by Age Group Highlighting Violence-Related Injury Deaths, United States—2011," in *Ten Leading Causes of Death and Injury*, Centers for Disease Control, 2014, http://www.cdc.gov/injury/wisqars/pdf/leading_causes_of_injury_death_highlighting_ violence_2011-a.pdf (accessed August 15, 2014)

would largely substantiate gun rights advocates' belief that firearm ownership makes the United States safer, rather than more dangerous. Few other gun researchers have produced estimates anywhere near as high as Kleck's, however, and Kleck's critics, such as David Hemenway of the Harvard Injury Control Research Center (HICRC), note many apparent methodological flaws in his research. Hemenway also argues that a comparison of Kleck's estimate with Federal Bureau of Investigation crime statistics and with data about the number of people admitted to hospitals for gunshot wounds shows the 2 million to 2.5 million number to be implausibly high. For his own part, Kleck considers Hemenway's critique to be ideologically motivated and methodologically dubious.

The Violence Policy Center (VPC), a nonprofit group that advocates for stricter gun control laws, is responsible for one of the most current estimates, as of November 2014, of how often guns are used in self-defense. The VPC observes in *Firearm Justifiable Homicides and Non-fatal Self-Defense Gun Use* (April 2013, http://www.vpc.org/studies/justifiable.pdf), which is based on analyses of Federal Bureau of Investigation and Bureau of Justice Statistics data for the five-year period between 2007 and 2011, that a total of 29.6 million people were the victims of attempted or completed criminal violence and that firearms were used defensively in only 0.8% of incidents. Thus, the VPC calculates a five-year total of 235,700 people, or an average of 47,140 people per year, who used firearms in self-defense.

In spite of the lack of objective, indisputable evidence of how often firearms are used in self-defense, scholars overwhelmingly support estimates closer to Hemenway's and the VPC's than to Kleck's. In a survey conducted in June 2014 (http://www.hsph.harvard.edu/wp-content/uploads/sites/1264/2014/05/Expert-Survey-2-Results.pdf), the HICRC asked experts in the fields of public health, public policy, sociology, and criminology to state their level of agreement with the statement "In the United States, guns are used in self-defense far more often than they are used in crime" and to rate the quality of the scientific evidence on the issue. Of 122 experts who completed the survey, 39% strongly disagreed with this statement, 34% disagreed, 11% neither agreed nor disagreed, 4% agreed, 4% strongly agreed, and 9% professed not to know. The experts' view of the quality of research on the topic was more mixed: 12% rated it very strong, 20% strong, 19% medium, 20% weak, 9% very weak, and 19% professed not to know.

Thus, 73% of experts believe with a high degree of certainty that guns are used in crime more often than they are used in self-defense, but only 32% believe there is strong evidence for such a position. This lack of overwhelming evidence is likely to remain a persistent feature of the debate over defensive gun use, given the difficulty of determining how many times guns are used in self-defense without being reported to law enforcement agencies.

Self-Defense Narratives

Regardless of the statistical balance between the use of guns for self-defense and the use of guns to commit homicide and suicide, the fact remains that many people do successfully use guns to defend themselves and their property. In the eyes of many gun rights advocates, these stories show that the right to self-defense is a vital one no matter the aggregate totals of defensive gun use. Accordingly, the NRA maintains a regularly updated feature on its website called "Armed Citizen" (http://www.nraila.org/gun-laws/armed-citizen.aspx) that summarizes local news accounts of law-abiding gun owners successfully defending themselves with guns. The organization posts two to four stories in a typical week, which demonstrates that such uses of firearms are indeed common by most definitions. For example, typical stories from August 2014 included an account of an employee in a Wisconsin optical store who shot and killed a man who was attempting to rob the store at gunpoint; an account of a man who thwarted a carjacking in his Phoenix, Arizona, neighborhood by wounding the thief with a handgun; and an account of an Albuquerque, New Mexico, man who was awoken by three armed robbers entering his home, retrieved his gun, killed one of the intruders, and wounded another.

EFFORTS TO PROMOTE GUN SAFETY
Safe Storage of Guns in the Home

If a gun is to be used for self-defense, does it make sense to keep it unloaded and locked up? This is a question asked by people who oppose safe-storage laws and laws that hold gun owners criminally liable for any injury caused by a child gaining unsupervised access to a gun. The Law Center to Prevent Gun Violence notes in "Child Access Prevention Policy Summary" (August 1, 2013, http://smartgunlaws.org/child-access-prevention-policy-summary) that as of 2013, 28 states and the District of Columbia had child access prevention (CAP) laws. Gun control advocates have long pressed for such legislation at the federal level.

Lisa Hepburn et al. conclude in "The Effect of Child Access Prevention Laws on Unintentional Child Firearm Fatalities, 1979–2000" (*Journal of Trauma Injury, Infection, and Critical Care*, vol. 61, no. 2, August 2006) that CAP laws may have influenced the continued reduction in unintentional firearm death rates that occurred between 1979 and 2000 nationally among children. The researchers determine that "the decrease in rates of unintentional firearm deaths for children aged 0 to 14 in CAP law states exceeded the average for states without CAP laws in 9 of the 14 states for which data were available." According to Hepburn et al., statistical analyses of the data "showed a significant association between CAP laws and rates of unintentional firearm deaths for children 0 to 14 years old."

Some argue that if society chooses to hold people accountable for negligent actions or child endangerment, it should do so with all such actions and not single out firearms. As Figure 6.3 shows, accidents (a category that includes drowning, falls, automobile accidents, fires, and many other events that occur both inside and outside of the home) accounted for 38% of the deaths of young people aged one to 24 years in 2011. Accidental shootings are extremely rare, accounting for only 591 total U.S. deaths in 2011. (See Table 6.8.) Thus, according to the line of reasoning put forward by those who oppose CAP laws, gun control advocates unfairly focus on guns when a variety of parental behaviors are more deadly than negligence having to do with guns.

CAP laws are intended not only to protect children from unintentional injury from firearms but also from gaining access to firearms to commit suicide and homicide. As Figure 6.3 shows, suicide and homicide each accounted for 13% of the deaths of young people in 2011, and as the CDC (January 9, 2014, http://www.cdc.gov/violenceprevention/pub/youth_suicide.html) notes, firearms are the leading method of suicide among young people (as among adults). However, the effectiveness of preventing youth suicide and homicide by restricting their access to guns has not been firmly established.

FIGURE 6.3

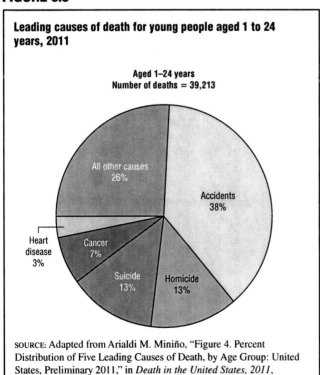

Leading causes of death for young people aged 1 to 24 years, 2011

Aged 1–24 years
Number of deaths = 39,213

All other causes 26%

Accidents 38%

Heart disease 3%

Cancer 7%

Suicide 13%

Homicide 13%

SOURCE: Adapted from Arialdi M. Miniño, "Figure 4. Percent Distribution of Five Leading Causes of Death, by Age Group: United States, Preliminary 2011," in *Death in the United States, 2011*, Centers for Disease Control and Prevention, National Center for Health Statistics, March 2013, http://www.cdc.gov/nchs/data/databriefs/db115.pdf (accessed July 16, 2014)

Jeffrey A. Bridge et al. report in "Changes in Suicide Rates by Hanging and/or Suffocation and Firearms among Young Persons Aged 10–24 Years in the United States: 1992–2006" (*Journal of Adolescent Health*, vol. 46, no. 5, May 2010) that between 1992 and 2006 youth suicides committed with firearms decreased; the researchers, however, do not know if this decrease was due to CAP laws. They note that suicide by poisoning and other methods declined simultaneously in this age group, while suicide by hanging and suffocation increased concurrently. These data suggest a selection of suffocation and hanging rather than a thwarting of other methods. The researchers indicate that hanging and suffocation were the leading methods of suicide among youth in many countries, such as England, Australia, and New Zealand, and recommend that future research investigate the reasons for this preference so that public health programs might be developed to curb this method of suicide.

Safety Programs to Protect Young Children

In 1988 the NRA created the Eddie Eagle GunSafe Program (http://www.nrahq.org/safety/eddie) for gun safety. Eddie Eagle is a school-based program that teaches gun safety to young children (preschool to third grade). With the help of the cartoon character Eddie the Eagle, kids are taught that when they find a gun they should not touch it, but instead leave the area and tell an adult. The VPC contends in "Joe Camel with Feathers: How the NRA with Gun and Tobacco Industry Dollars Uses Its Eddie Eagle Program to Market Guns to Kids" (November 1997, http://www.vpc.org/studies/eddiecon.htm) that the program is a marketing tool for the NRA, allowing it to whitewash the dangers of guns and recruit a new generation of customers for the firearms industry. A segment on the ABC News program *20/20*, which aired in 1999, made a similar argument. Michael B. Himle et al. assess in "An Evaluation of Two Procedures for Training Skills to Prevent Gun Play in Children" (*Pediatrics*, vol. 113, no. 1, January 1, 2004) the effectiveness of the Eddie Eagle program and a separate program of behavioral skills training. They find that both programs increased children's awareness of the tenets of gun safety, that the behavioral training program alone (and not the Eddie Eagle program) was effective in teaching gun-safety skills as observed during supervised role play, and that neither program was effective in teaching children to apply gun safety skills outside of the controlled environment of the training sessions.

In "Peer Tutoring to Prevent Firearm Play: Acquisition, Generalization, and Long-Term Maintenance of Safety Skills" (*Journal of Applied Behavior Analysis*, vol. 41, no. 1, 2008), Candace M. Jostad et al. support the behavioral skills training approach and show that six- and seven-year-old children acquire and maintain firearm safety skills when behavioral skills training is used and is taught by other children trained in the techniques.

National Goals: *Healthy People 2020*

Released in December 2010, *Healthy People 2020* (http://www.healthypeople.gov) is a set of national health objectives developed under the leadership of a variety of U.S. governmental agencies, such as the CDC; the Health Resources and Services Administration; the President's Council on Sports, Fitness, and Nutrition; and the U.S. Food and Drug Administration. This national health initiative builds on previous programs that began with *Healthy People: The Surgeon General's Report on Health Promotion and Disease Prevention* (1979, http://profiles .nlm.nih.gov/NN/B/B/G/K) and have been updated and reassessed in each decade since. *Healthy People 2020* replaces the previous initiative, *Healthy People 2010* (January 2000, http://www.cdc.gov/nchs/healthy_people/ hp2010.htm).

The *Healthy People 2020* objectives with regard to firearms are included in the section outlining Injury and Violence Prevention (IVP) goals:

- IVP-30: To reduce the rate of firearm-related deaths 10%, from 10.3 firearm-related deaths per 100,000 population in 2007 to 9.3 deaths per 100,000 population in 2020

- IVP-31: To reduce the rate of nonfatal firearm-related injuries by 10%, from 20.7 nonfatal firearm-related injuries per 100,000 population in 2007 to 18.6 injuries per 100,000 population in 2020

As of 2014, no progress had been made toward reducing either the firearm-related death rate or the nonfatal firearm-related injury rate, both of which had fluctuated slightly in the years since 2007 while remaining flat overall.

CHAPTER 7

GUNS AND YOUTH

Young people are often at the center of the gun control debate. Gun rights advocates maintain that responsible firearm use is the key to gun safety, but children, adolescents, and many young adults lack the capacity for sound judgment that is the necessary pre-condition to responsible firearm use. A patchwork of federal and state laws is designed to prevent the purchase and possession of firearms by minors (in some cases defined as those under the age of 21 years and in other cases as those under the age of 18 years) except under special circumstances, but young people who commit murder or suicide often do so with guns owned by their parents or family members, and guns used by underage criminals are often obtained illegally. The question of how to balance gun owners' rights with the public welfare thus inevitably arises when young people kill one another or themselves with firearms.

During the late 1980s and early 1990s youth gang activity in urban areas appeared to be permanently altering the nature of city life, and school shootings began occurring and then reoccurring with ominous regularity. Together, these trends were among gun control advocates' most powerful reasons to push for stricter gun laws. However, during the late 1990s and early 2000s the homicide and crime rates nationwide and in many previously violent urban areas declined dramatically. The reasons for the decline in crime rates have not been conclusively established, but little evidence exists to suggest that it was a result of gun control efforts because the period was not marked by any substantial shift in gun laws.

Youth firearm violence remains a major concern in many urban areas, however, and mass shootings in schools and other public areas have by most accounts increased in regularity since the first decade of the 21st century. Additionally, the terror inspired by mass shootings reached new heights in 2012, when 20 first graders were killed by a mentally unstable 20-year-old at Sandy Hook Elementary School in Newtown, Connecticut. Finally, as discussed in Chapter 6, although the firearm homicide rate has declined nationally, the firearm suicide rate remains high. The suicide rate for the U.S. population has neither increased nor decreased dramatically since the mid-20th century, but the suicide rate for young people, and for young males in particular, has increased significantly since that time.

DEADLY ASSAULTS

In *Homicide Trends in the United States, 1980–2008* (November 2011, http://bjs.ojp.usdoj.gov/content/pub/pdf/htus8008.pdf), Alexia Cooper and Erica L. Smith of the Bureau of Justice Statistics (BJS) attribute the increase in homicides during the late 1980s and early 1990s and the subsequent decline to gun violence among teenagers and young adults; homicide victimization rates for teens (aged 14 to 17 years) peaked in 1993 at 12 homicides per 100,000 and declined after that time to 5.1 homicides per 100,000 in 2008. Similarly, homicide offending reached a high of 30.7 teen offenders per 100,000 in that age group in 1993; since 2000 the rate of homicide offenders among teenagers has remained relatively stable at a level similar to that which occurred in 1985 (9.5 per 100,000).

Figure 7.1 shows homicide trends between 1980 and 2008 by victim/offender relationship and weapon use. The rise in gun-related homicides between 1984 and 1994 is evident in homicides by friends and acquaintances and is even more pronounced in homicides by strangers. Gun violence among family members and those in intimate relationships showed a gradual decline during the 1980s and 1990s. Among family members, other weapons (a category that includes knives, sharp instruments, and blunt objects as well as personal weapons such as hands and feet) overtook guns as the most likely type of weapon to be used in a homicide.

FIGURE 7.1

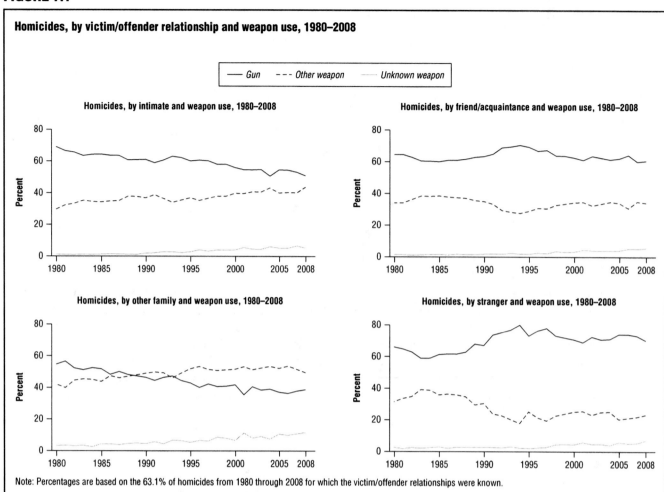

Homicides, by victim/offender relationship and weapon use, 1980–2008

— Gun   – – – Other weapon   ·········· Unknown weapon

Homicides, by intimate and weapon use, 1980–2008

Homicides, by friend/acquaintance and weapon use, 1980–2008

Homicides, by other family and weapon use, 1980–2008

Homicides, by stranger and weapon use, 1980–2008

Note: Percentages are based on the 63.1% of homicides from 1980 through 2008 for which the victim/offender relationships were known.

SOURCE: Adapted from Alexia Cooper and Erica L. Smith, "Figure 25a. Homicides, by Intimate and Weapon Use, 1980–2008," "Figure 25b. Homicides, by Other Family and Weapon Use, 1980–2008," "Figure 25c. Homicides, by Friend/Acquaintance and Weapon Use, 1980–2008," and "Figure 25d. Homicides, by Stranger and Weapon Use, 1980–2008," in *Homicide Trends in the United States, 1980–2008*, U.S. Department of Justice, Bureau of Justice Statistics, November 2011, http://bjs.ojp.usdoj.gov/content/pub/pdf/htus8008.pdf (accessed July 17, 2014)

Table 7.1 presents the number and rate of firearm deaths for those aged zero to 19 years in 2011. The highest victimization rate occurred among male teenagers aged 15 to 19 years at 17.67 firearm deaths per 100,000 population. Females in this same age group registered the highest rate among girls, at 2.42 per 100,000. With the exception of boys aged 10 to 14 years, whose firearm death rate was slightly elevated, at 1.49 per 100,000, children under the age of 14 years are among the least likely of all population groups to be killed by firearms.

Youthful Offenders

Figure 7.2 depicts trends in homicide offenders by age group between 1980 and 2008. As the three graphs show, homicides involving weapons other than guns declined slightly for all age groups during this period, with few significant increases or decreases from year to year. The same general pattern held for gun homicides committed by offenders aged 25 years and older. However, firearm

TABLE 7.1

Firearm deaths of children aged 0 to 19 years and rates per 100,000, 2011

Age group	Sex	Deaths	Population	Rate
00–04	Males	32	10,281,514	0.31
	Females	23	9,846,375	0.23
		55	20,127,889	0.27
05–09	Males	32	10,384,932	0.31
	Females	24	9,950,826	0.24
		56	20,335,758	0.28
10–14	Males	158	10,590,305	1.49
	Females	41	10,120,886	0.41
		199	20,711,191	0.96
15–19	Males	1,965	11,117,723	17.67
	Females	255	10,530,679	2.42
		2,220	21,648,402	10.25
Total		**2,530**	**82,823,240**	**3.05**

SOURCE: Adapted from "Fatal Injury Reports, National and Regional, 1999–2011," in *Web-Based Injury Statistics Query and Reporting System (WISQARS)*, Centers for Disease Control and Prevention, 2014, http://webappa.cdc.gov/sasweb/ncipc/mortrate10_us.html (accessed July 16, 2014)

FIGURE 7.2

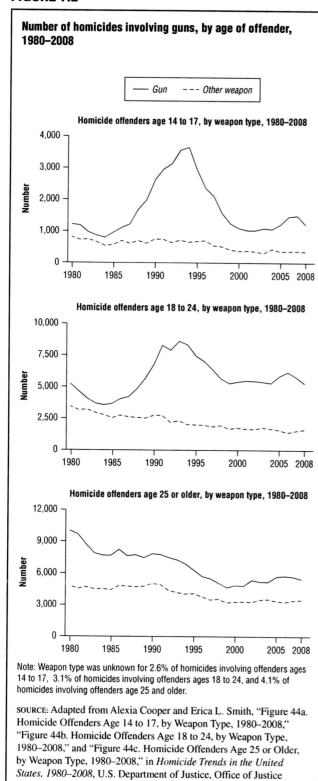

Number of homicides involving guns, by age of offender, 1980–2008

Gun —— Other weapon - - -

Homicide offenders age 14 to 17, by weapon type, 1980–2008

Homicide offenders age 18 to 24, by weapon type, 1980–2008

Homicide offenders age 25 or older, by weapon type, 1980–2008

Note: Weapon type was unknown for 2.6% of homicides involving offenders ages 14 to 17, 3.1% of homicides involving offenders ages 18 to 24, and 4.1% of homicides involving offenders age 25 and older.

SOURCE: Adapted from Alexia Cooper and Erica L. Smith, "Figure 44a. Homicide Offenders Age 14 to 17, by Weapon Type, 1980–2008," "Figure 44b. Homicide Offenders Age 18 to 24, by Weapon Type, 1980–2008," and "Figure 44c. Homicide Offenders Age 25 or Older, by Weapon Type, 1980–2008," in *Homicide Trends in the United States, 1980–2008*, U.S. Department of Justice, Office of Justice Programs, Bureau of Justice Statistics, November 2011, http://bjs.ojp.usdoj.gov/content/pub/pdf/htus8008.pdf (accessed July 17, 2014)

peaked during the mid-1990s and then declined almost as dramatically as it had risen. A similar pattern holds when limiting one's view to juvenile offenders (those under the age of 18 years, with the exception of a small number of under-18 offenders who are tried as adults). Figure 7.3 shows the number of juveniles who committed homicide with and without guns between 1980 and 2011. The graph shows a dramatic rise in the number of juveniles who committed homicide with a gun beginning in 1984 and peaking in 1994.

Thus, the increase in the national homicide rate during the 1980s and 1990s was largely an increase in firearm homicides committed by teenagers and young adults, and the decrease in firearm homicides since that time was largely a decrease among this same age group. Howard N. Snyder and Melissa Sickmund of the National Center for Juvenile Justice provide in *Juvenile Offenders and Victims: 2006 National Report* (March 2006, http://www.ojjdp.gov/ojstatbb/nr2006/downloads/NR2006.pdf) further detail regarding the 1984 to 1994 rise. According to the researchers, 90% of the overall increase was attributable to males killing nonfamily members with guns, usually handguns. Snyder and Sickmund state, "This type of murder increased 400% between 1984 and 1994. A closer look at these crimes reveals that the increase was somewhat greater for murders of acquaintances than strangers and somewhat greater for juveniles acting with other offenders than for a juvenile offender acting alone. Nearly three-quarters of the increase was the result of crimes committed by black and other minority males—and in two-thirds of these murders, the victims were minority males."

Why did gun-related homicides rise between 1984 and 1994? The Virginia Youth Violence Project at the University of Virginia's Curry School of Education suggests in "Juvenile Homicide" (2010) that violent juvenile crime during that time was linked to many factors, including the introduction of crack cocaine, the availability of cheap handguns, inadequate after-school supervision, and the prevalence of violence in the media. Reasons for the subsequent fall have been much debated, and experts disagree about the relative importance of different factors. Some of the most widely acknowledged factors explaining the trend toward improved safety in urban areas include increases in the number of police as well as the introduction of new police techniques; the gentrification of urban areas, whereby wealthier residents moved into previously impoverished neighborhoods and businesses opened to serve their needs, leading to increasing real estate prices and the dispersal of criminal activity; and the receding crack epidemic.

JUVENILE ARRESTS. Table 7.2 shows the estimated number of juvenile arrests in 2011 and the percentage change in juvenile arrests over various time spans. Juvenile

homicides committed by teens and adults aged 14 to 24 years followed neither of these patterns. Instead, youth offenders committed firearm homicides at a dramatically increasing rate during the late 1980s and early 1990s. The rate at which young people committed firearm homicides

FIGURE 7.3

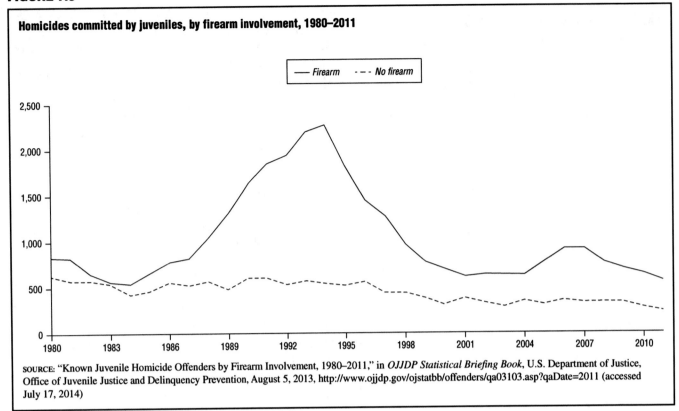

Homicides committed by juveniles, by firearm involvement, 1980–2011

—— Firearm - - - No firearm

SOURCE: "Known Juvenile Homicide Offenders by Firearm Involvement, 1980–2011," in *OJJDP Statistical Briefing Book*, U.S. Department of Justice, Office of Juvenile Justice and Delinquency Prevention, August 5, 2013, http://www.ojjdp.gov/ojstatbb/offenders/qa03103.asp?qaDate=2011 (accessed July 17, 2014)

crime fell 31% overall between 2002 and 2011, and violent crime perpetrated by juveniles fell 27%. Decreases during this period were significant in most types of crimes in which firearms were typically used, including murder and nonnegligent manslaughter (−36%), forcible rape (−40%), and aggravated assault (−34%). Juvenile rates of robbery, another crime in which firearms are often used, decreased by a comparatively modest 3%. Meanwhile, weapons offenses fell 20% over the course of those 10 years.

RACE AND GENDER. Young males, and young African American males in particular, are disproportionately likely to be homicide offenders. As Figure 7.4 shows, between 1980 and 2008 white males aged 14 to 24 years represented between 9% and 6% of the total population, while accounting for more than 15% of homicide offenders. Meanwhile, African American males of the same age group made up around 1% of the total population and accounted for between 17% and 34% of homicide offenders. The share of young African American males among all homicide offenders increased dramatically from the late 1980s through the mid-1990s, the years when the homicide rate itself increased. African American males aged 14 to 24 years continued to be comparably overrepresented among homicide offenders in the 21st century.

Youthful Victims

Young people are often the victims of violent crimes, in many cases because they have been targeted by other young people. In "Age 14 Starts a Child's Increased Risk of Major Knife or Gun Injury in Washington, D.C." (*Journal of the National Medical Association*, vol. 96, no. 2, February 2004), Howard A. Freed et al. analyze trauma registry data at an inner-city trauma center over a period of eight years. The researchers find that the risk of a youth becoming a victim of a major gunshot wound or stabbing rose dramatically at age 14 and that this risk continued to rise sharply through age 18.

Figure 7.5 shows that between 1980 and 2008 more than three-quarters of homicide victims aged 16 to 22 years were killed with guns and that 17 was the peak age for homicide by firearm (79% of 17-year-old homicide victims were killed with guns). For purposes of comparison, note that less than half of homicide victims aged 63 years and older or aged 11 years and younger were killed with guns during the same period.

Young children are less frequently the victims of gun violence than older children. Table 7.3 shows that both the number and the rate of homicide firearm deaths for children aged 12 years and younger remained steady between 2002 and 2011. The number and rate of firearm homicide deaths in this age group varied from a low of 121 deaths and a rate of 0.23 per 100,000 in 2004 to a high of 175 deaths and a rate of 0.34 in 2002. There were fluctuations within this range but no general observable tendency in either direction over the course of the decade.

TABLE 7.2

Juvenile arrests, 2011, and percentage change, 2002–11

The number of arrests of juveniles in 2011 was 31% fewer than the number of arrests in 2002

Most serious offense	2011 estimated number of juvenile arrests	Percent of total juvenile arrests			Percent change		
		Female	Younger than 15	White	2002–2011	2007–2011	2010–2011
Total	1,470,000	29%	27%	66%	–31%	–29%	–11%
Violent crime index	68,150	18	27	47	–27	–29	–10
Murder and nonnegligent manslaughter	840	9	11	45	–36	–37	–17
Forcible rape	2,800	2	35	64	–40	–22	–2
Robbery	23,800	9	19	30	–3	–31	–13
Aggravated assault	40,700	25	32	56	–34	–29	–9
Property crime index	334,700	36	28	62	–30	–20	–9
Burglary	62,000	12	27	60	–28	–24	–5
Larceny-theft	253,800	44	28	63	–25	–15	–10
Motor vehicle theft	14,000	16	20	55	–69	–52	–11
Arson	4,900	15	57	72	–40	–31	8
Nonindex							
Other (simple) assaults	190,900	36	39	59	–19	–21	–9
Forgery and counterfeiting	1,600	29	13	66	–69	–49	–7
Fraud	5,200	35	16	60	–39	–27	–9
Embezzlement	400	39	10	66	–70	–75	–6
Stolen property (buying, receiving, possessing)	13,300	17	22	57	–48	–40	–9
Vandalism	67,900	15	39	76	–35	–39	–12
Weapons (carrying, possessing, etc.)	28,200	10	33	61	–20	–35	–10
Prostitution and commercialized vice	1,000	76	9	35	–32	–34	–7
Sex offense (except forcible rape and prostitution)	12,600	11	49	70	–35	–19	–3
Drug abuse violations	148,700	17	17	74	–20	–24	–13
Gambling	1,000	6	11	12	–35	–52	–28
Offenses against the family and children	3,600	39	31	70	–62	–39	–6
Driving under the influence	10,100	25	2	91	–53	–44	–16
Liquor laws	88,300	40	9	89	–39	–37	–7
Drunkenness	11,400	26	12	87	–39	–32	–10
Disorderly conduct	139,200	35	38	57	–28	–31	–11
Vagrancy	1,800	23	29	73	–10	–53	–15
All other offenses (except traffic)	264,900	26	23	69	–33	–29	–11
Suspicion (not included in totals)	100	28	31	51	–90	–65	8
Curfew and loitering	76,900	30	25	61	–46	–46	–19

Notes: All four offenses that make up the violent crime index decreased considerably between 2007 and 2011: murder (–37%), rape (–22%), robbery (–31%), and aggravated assault (–29%). In 2011, there were an estimated 190,900 juvenile arrests for simple assault. Approximately one-third (36%) of these arrests involved females, 39% involved youth younger than age 15, and 59% involved white youth. Youth younger than age 15 accounted for more than half (57%) of all juvenile arrests for arson in 2011 and nearly 40% of juvenile arrests for simple assault, vandalism, and disorderly conduct. Females accounted for 9% of juvenile arrests for murder but 25% of juvenile arrests for aggravated assault. Detail may not add to totals because of rounding.

SOURCE: Charles Puzzanchera, "The Number of Arrests of Juveniles in 2011 Was 31% Fewer Than the Number of Arrests in 2002," in *Juvenile Arrests 2011*, U.S. Department of Justice, Office of Juvenile Justice and Delinquency Prevention, December 2013, http://www.ojjdp.gov/pubs/244476.pdf (accessed July 17, 2014)

RACE AND GENDER. Besides being disproportionately likely to be homicide offenders, young males, and young African American males in particular, are disproportionately likely to be homicide victims. As noted earlier, between 1980 and 2008 the proportion of white males aged 14 to 24 years among the total U.S. population ranged from a high of 9% to a low of 6% and African American males of the same age group accounted for approximately 1% of the total population. (See Figure 7.4.) During this period young white males accounted for approximately 10% of homicide victims and young African American males accounted for between 9% and 18% of all homicide victims.

Smith and Cooper note in *Homicide in the U.S. Known to Law Enforcement, 2011* (December 2013, http://www.bjs.gov/content/pub/pdf/hus11.pdf) the homicide rate for both white and African American males increases sharply at age 14 and peaks during the early 20s, but the magnitude of the increase differs dramatically for each group. Between 2002 and 2011 the homicide rate for white males peaked at age 20 at 11.4 per 100,000, while the homicide rate for African American males peaked at age 23 at 100.3 per 100,000. Thus, even after the dramatic declines in homicide rates during the late 1990s, young African American males had a peak homicide rate almost nine times greater than that of young white males.

Smith and Cooper further observe that more than 75% of young white male homicide victims were killed with firearms between 2008 and 2011, slightly down

FIGURE 7.4

Males aged 14 to 24 as a proportion of homicide offenders, homicide victims, and the population, by race, 1980–2008

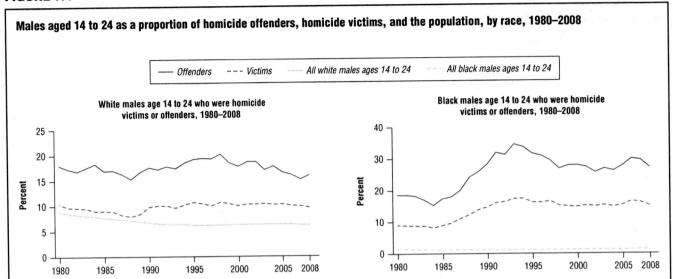

SOURCE: Adapted from Alexia Cooper and Erica L. Smith, "Figure 23a. White Males Age 14 to 24 Who Were Homicide Victims or Offenders, 1980–2008," and "Figure 23b. Black Males Age 14 to 24 Who Were Homicide Victims or Offenders, 1980–2008," in *Homicide Trends in the United States, 1980–2008*, U.S. Department of Justice, Bureau of Justice Statistics, November 2011, http://bjs.ojp.usdoj.gov/content/pub/pdf/htus8008.pdf (accessed August 16, 2014)

FIGURE 7.5

Percent of homicides involving guns, by age of victim, 1980–2008

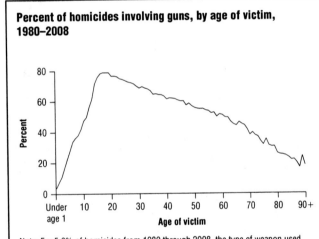

Note: For 5.0% of homicides from 1980 through 2008, the type of weapon used was unknown.

SOURCE: Alexia Cooper and Erica L. Smith, "Figure 43. Homicides Involving Guns, by Age of Victim, 1980–2008," in *Homicide Trends in the United States, 1980–2008*, U.S. Department of Justice, Office of Justice Programs, Bureau of Justice Statistics, November 2011, http://bjs.ojp.usdoj.gov/content/pub/pdf/htus8008.pdf (accessed August 28, 2012)

TABLE 7.3

Homicide firearm deaths of children aged 12 and under and rate per 100,000, 2002–2011

Year	Deaths	Population	Crude rate
2002	175	52,221,496	0.34
2003	146	52,005,373	0.28
2004	121	51,893,344	0.23
2005	126	51,856,979	0.24
2006	160	51,943,526	0.31
2007	156	52,214,785	0.30
2008	161	52,509,274	0.31
2009	157	52,737,288	0.30
2010	142	52,943,218	0.27
2011	145	52,920,466	0.27
Total	**1,489**	**523,245,749**	**0.28**

SOURCE: Adapted from "Fatal Injury Reports, National and Regional, 1999–2011," in *Web-Based Injury Statistics Query and Reporting System (WISQARS)*, Centers for Disease Control and Prevention, 2014, http://webappa.cdc.gov/sasweb/ncipc/mortrate10_us.html (accessed July 17, 2014)

from the more than 80% killed with firearms between 1992 and 1995, the period when the homicide rate peaked. (See Figure 7.6.) Meanwhile, over 90% of young African American male homicide victims were killed with firearms between 2008 and 2011, a proportion that was unchanged from the period of peak homicide rates. (See Figure 7.7.)

According to Smith and Cooper, between 2002 and 2011 young white females were less likely to be homicide victims than young white males, and young African American females were less likely to be homicide victims than young African American males. For young white females, the homicide rate peaked in infancy: girls younger than one had a homicide rate of 4.5 per 100,000. African American girls under age one also had a notably high homicide rate of 10.3 per 100,000, but the peak rate for young African American females came at age 22, at 11.8 per 100,000. Thus, although young males in general were far more likely to be homicide victims than young females, African American females aged 15 to 24 years were approximately as likely as white males of the same age to be homicide victims.

FIGURE 7.6

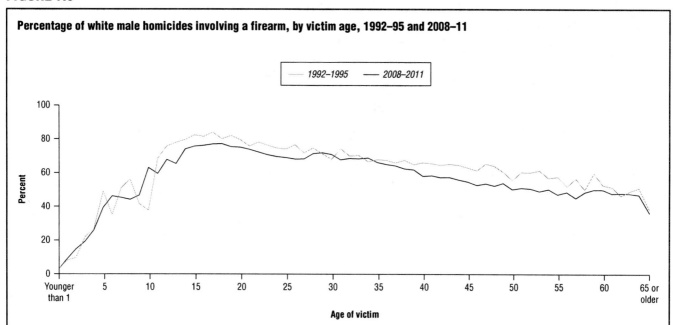

Percentage of white male homicides involving a firearm, by victim age, 1992–95 and 2008–11

— 1992–1995 — 2008–2011

Percent

100

80

60

40

20

0

Younger than 1   5   10   15   20   25   30   35   40   45   50   55   60   65 or older

Age of victim

SOURCE: Erica L. Smith and Alexia Cooper, "Figure 13. Percent of White Male Homicides Involving a Firearm, by Victim Age, 1992–95 and 2008–11," in *Homicide in the U.S. Known to Law Enforcement, 2011*, U.S. Department of Justice, Bureau of Justice Statistics, December 2013, http://www.bjs.gov/content/pub/pdf/hus11.pdf (accessed July 15, 2014)

FIGURE 7.7

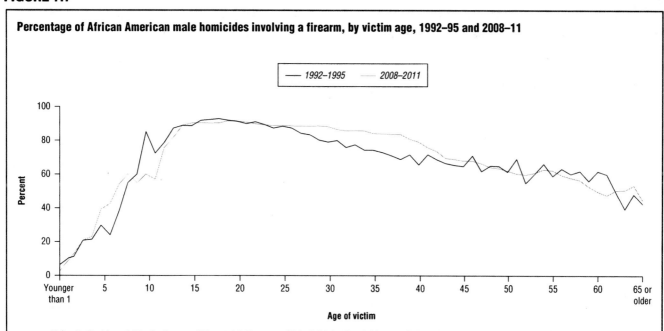

Percentage of African American male homicides involving a firearm, by victim age, 1992–95 and 2008–11

— 1992–1995 — 2008–2011

Percent

100

80

60

40

20

0

Younger than 1   5   10   15   20   25   30   35   40   45   50   55   60   65 or older

Age of victim

SOURCE: Erica L. Smith and Alexia Cooper, "Figure 15. Percent of Black Male Homicides Involving a Firearm, by Victim Age, 1992–95 and 2008–11," in *Homicide in the U.S. Known to Law Enforcement, 2011*, U.S. Department of Justice, Bureau of Justice Statistics, December 2013, http://www.bjs.gov/content/pub/pdf/hus11.pdf (accessed July 15, 2014)

Additionally, young African American female homicide victims were more likely than young white female victims to be killed with guns. Among young white females, the percentage of homicides involving guns peaked at 60% at around age 15 between 2008 and 2011, little different from observable trends during the peak homicide rate between 1992 and 1995. (See Figure 7.8.) Among young

African American females between 2008 and 2011, the percentage of homicides involving guns peaked at the same age but at a much higher level of around 75%. (See Figure 7.9.) Furthermore, among African American female homicide victims in their late teens and mid-20s, the percentage who were killed with guns had risen significantly from the peak homicide period between 1992 and 1995.

FIGURE 7.8

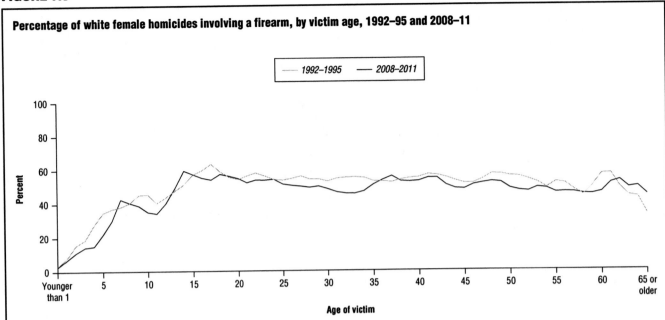

Percentage of white female homicides involving a firearm, by victim age, 1992–95 and 2008–11

SOURCE: Erica L. Smith and Alexia Cooper, "Figure 14. Percent of White Female Homicides Involving a Firearm, by Victim Age, 1992–95 and 2008–11," in *Homicide in the U.S. Known to Law Enforcement, 2011*, U.S. Department of Justice, Bureau of Justice Statistics, December 2013, http://www.bjs.gov/content/pub/pdf/hus11.pdf (accessed July 15, 2014)

FIGURE 7.9

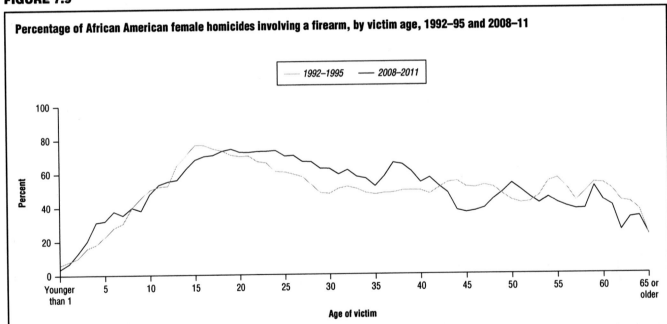

Percentage of African American female homicides involving a firearm, by victim age, 1992–95 and 2008–11

SOURCE: Erica L. Smith and Alexia Cooper, "Figure 16. Percent of Black Female Homicides Involving a Firearm, by Victim Age, 1992–95 and 2008–11," in *Homicide in the U.S. Known to Law Enforcement, 2011*, U.S. Department of Justice, Bureau of Justice Statistics, December 2013, http://www.bjs.gov/content/pub/pdf/hus11.pdf (accessed July 15, 2014)

YOUNG, ARMED, AND DANGEROUS

Youth Gangs

Much of the violent activity among young people can be attributed to youth gangs, which tend to be concentrated in poor, inner-city neighborhoods. The Office of Juvenile Justice and Delinquency Prevention's National Youth Gang Center has surveyed law enforcement agencies annually since 1996 to determine the extent of gang problems nationwide and to investigate demographic and other gang-related data. The ongoing National Youth Gang Surveys (NYGS; http://www.nationalgangcenter.gov/Survey-Analysis) reveal that the prevalence of gang problems nationwide declined from 1996 to 2001 but rose again through 2012. The survey found that there were more than 30,000 gangs in the United States in 2012, the highest annual estimate since 1996 (and up 15% since 2006). The number of individual gang members was also on the rise: in 2012 there were an estimated 850,000 gang members nationally, an 8.6% increase from 2011.

Over time, the results of the NYGS have consistently indicated that gang activity is concentrated in metropolitan areas (areas with an urban core and suburbs whose populations exceed 100,000), and especially in the larger cities themselves (cities of over 100,000 people). In 2012 law enforcement agencies in 3,100 U.S. jurisdictions, or 30% of all responding jurisdictions, reported gang problems in 2012. The percentage of metropolitan jurisdictions reporting gang activity was considerably higher, at 86%. Overall, 67.4% of all gangs and 81.7% of all gang members were located in metropolitan areas, and these urban/suburban areas accounted for 84.5% of all gang-related homicides. There were 2,363 gang homicides in 2012, up significantly from 2011, when there were 1,824 gang homicides. Over two-thirds (67.2%) of these homicides occurred in cities with populations of more than 100,000, and another 17.3% occurred in suburban areas. Smaller cities (those with populations between 50,000 and 100,000) and rural areas together accounted for 15.6% of all gang homicides in 2012.

The age makeup of gangs has changed meaningfully since the NYGS began in the mid-1990s, when about half the known gang members were juveniles. In 1996, 50% of gang members were juveniles and 50% were over the age of 18 years, but in every survey since then, juveniles have accounted for a minority of gang members. In 2011 roughly 35% of gang members were juveniles and 65% were over the age of 18 years. Gang membership tends to be older in more populous cities and counties in which gangs are well established, whereas smaller cities and rural areas, where the development of gang problems may be more recent, are more likely to report a higher proportion of juvenile gang members. Over 90% of gang members have been male in all years covered by the survey.

Criminally Active Nongang Groups

Gangs are not always the source of gun violence, however. Anthony A. Braga of the Malcolm Wiener Center for Social Policy at Harvard University reports in *Gun Violence among Serious Young Offenders* (March 29, 2004, http://www.cops.usdoj.gov/pdf/pop/e01042199.pdf) that in some cities criminally active nongang groups are major gun offenders.

Braga explains that gun violence and murders in gangs are usually related to rivalries among gangs; offenders often become victims and vice versa. Gun violence and murders in criminally active groups that are not gangs are usually related to "business interests," such as drug dealing. Murders tied to these groups usually occur in or near "street" drug markets, and many of the victims are part of the drug organization or criminal network.

In "Gun Carrying and Drug Selling among Young Incarcerated Men and Women" (*Journal of Urban Health*, vol. 83, no. 2, March 2006), Deborah Kacanek and David Hemenway study the correlation between illegal drug dealing and guns. Based on interviews with 204 state prison inmates between the ages of 18 and 25 years, the researchers find that 45% of the incarcerated men and 16% of the women reported carrying a weapon in the 12 months prior to their arrest. Gun-ownership was more prevalent among the men (55%) than the women (16%). Two-thirds of the men (67%) and 28% of the women reported that they had been shot at. Of those who were shot at, 16% of the men and 6% of the women had been wounded. Kacanek and Hemenway conclude that survey participants who sold drugs were more likely to have carried a gun: 65% of men and 22% of women who sold crack cocaine reported carrying a gun, as did 49% of men and 27% of women who sold drugs but not crack, compared with just 16% of men and 3% of women who did not sell drugs at all.

How Do Young Offenders Acquire Guns?

GUN TRACE DATA. As discussed in Chapter 2, the Bureau of Alcohol, Tobacco, Firearms, and Explosives (ATF) has been prevented by law from publicly releasing certain categories of gun trace statistics since 2003. The ATF currently releases state-by-state data on the total number of guns recovered by law enforcement agencies, but it is prevented from releasing crucial data regarding how those guns were obtained. This has hampered research into the understanding of how young people obtain guns, and the most informative national studies currently available date from before that time. The ATF's Youth Crime Gun Interdiction Initiative, which analyzed guns recovered from crimes and traced them to their original sources, was the foundation for much of the valuable gun trace analysis that existed in the public domain prior to 2003. According to the ATF, in *Crime Gun Trace Reports (2000): National Report* (July 2002), in 2000 there were a total of 88,570 crime firearms trace reports from 46 participating cities with populations exceeding 250,000. About 8% of crime guns were recovered from juveniles younger than 17 years old. About 33% of crime guns were recovered from young people between the ages of 18 and 24 years.

The ATF found that many recovered firearms moved rapidly from retail sales at federally licensed gun dealers to the black market (a market where products are bought and sold illegally), which supplies juveniles with guns. When crime guns were recovered within three years from the time of sale, they were more easily traced to their illegal sources than older guns, which were more likely to have passed through many hands before entering the illegal market. According to the ATF, these "new" crime guns made up nearly one-third of all firearms recovered in 2000.

Garen J. Wintemute et al. analyze in "The Life Cycle of Crime Guns: A Description Based on Guns Recovered from Young People in California" (*Annals of Emergency Medicine*, vol. 43, no. 6, June 2004) data from ATF firearms tracing records to follow the life cycle of 2,121 crime guns recovered in California in 1999. The researchers make several interesting conclusions:

- Guns recovered from individuals younger than 18 years old were most often purchased by people aged 45 years and older.

- Small-caliber handguns made up 41% of handguns recovered from this group.

- For 17.3% of crime guns recovered from teenagers, the median time from sale to recovery was less than three years, which indicates deliberate gun trafficking (the illegal selling of guns and ammunition).

- A minority of retailers and straw purchasers (people buying guns for someone else) are disproportionately linked to the sale or transfer of crime guns.

In the 21st century Chicago, Illinois, had one of the worst gang violence problems in the United States, and its murder rate was substantially higher than that of other large cities, including New York City and Los Angeles. In *Tracing the Guns: The Impact of Illegal Guns on Violence in Chicago* (May 27, 2014, https://www.cityof chicago.org/content/dam/city/depts/mayor/Press%20Room/ Press%20Releases/2014/May/05.27.14TracingGuns.pdf), the Chicago Office of the Mayor and the Chicago Police Department report on the findings of gun traces conducted on firearms used in crimes between 2009 and 2013. Of those guns whose purchase records could be traced, approximately 60% originated in states other than Illinois and were illegally trafficked into the state. The largest out-of-state sources for guns used in Chicago crimes were Indiana (which supplied 19% of the guns used in crimes), Mississippi (6.7%), and Wisconsin (3.6%), none of which require background checks for gun purchases at gun shows or on the Internet. The remaining 40% of guns used in Chicago crimes were purchased in Illinois, in spite of the state's stricter gun control laws. However, four Chicago-area dealers legally sold nearly 20% of all guns used in Chicago crimes, a fact that suggests that these dealers may have tacitly allowed for routine straw purchases.

OTHER STUDIES. Other studies have used research resources other than gun trace data to determine how young people who use guns in crimes obtain their weapons. Daniel W. Webster et al. of Johns Hopkins University interviewed 45 youths incarcerated in a juvenile justice facility to determine how they obtained guns. The researchers reported their findings in "How Delinquent Youths Acquire Guns: Initial versus Most Recent Gun Acquisitions" (*Journal of Urban Health*, vol. 79, no. 1, March 2002). Of the 45 youths, 30 had acquired at least one gun and 22 had acquired multiple guns. Approximately 50% of their first guns were given to them by friends or family, or they found discarded guns. Those who acquired more than one gun usually got them from acquaintances or drug addicts. If they bought new guns, the youths generally purchased them from gun traffickers (people who are in the business of selling guns illegally). Webster et al. conclude that a way to reduce the number of guns in the hands of young offenders is to stop high-volume gun traffickers and recover discarded guns from areas in which illicit drug sales take place.

In "Source of Firearms Used by Students in School-Associated Violent Deaths—United States, 1992–1999" (*Morbidity and Mortality Weekly Report*, vol. 52, no. 9, March 7, 2003), the Centers for Disease Control and Prevention (CDC) investigates how students obtained firearms used in serious school-associated crimes such as homicide and suicide. Most students obtained guns from home (37.5%), with the next likely source being a friend or relative (23.4%). Only 7% of guns used in school-related crimes were purchased and 5.5% were stolen.

Students and Guns

Since 1991 the CDC has regularly surveyed young people as part of its Youth Risk Behavior Surveillance System, which collects information about dangerous behaviors and risk factors among school-aged young people. Figure 7.10 shows the percentage of students by gender in grades nine to 12 who reported carrying a weapon anywhere and on school property between 1993 and 2011. The prevalence of weapon-carrying declined considerably both away from school (anywhere) and on school property over this period. Among males, the proportion who reported carrying a weapon anywhere fell from 34.3% in 1993 to 25.9% in 2011, and the proportion who reported carrying a weapon at school fell from 17.9% to 8.2%. The proportion of females who carried weapons either away from school or at school was much lower throughout these years and showed only a slight downward trend.

Table 7.4 breaks down the prevalence of weapon-carrying among high school students by sex, race, ethnicity, grade, and location. Native American or Alaskan

FIGURE 7.10

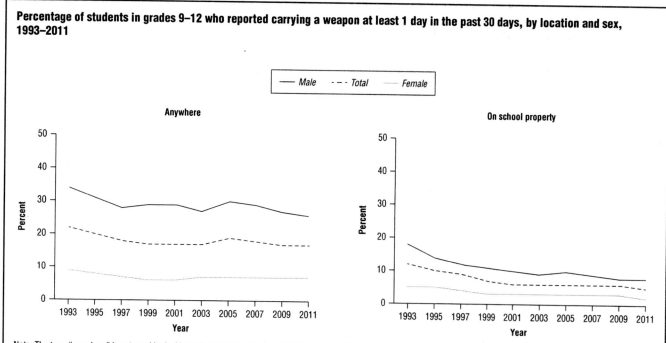

Percentage of students in grades 9–12 who reported carrying a weapon at least 1 day in the past 30 days, by location and sex, 1993–2011

Note: The term "anywhere" is not used in the Youth Risk Behavior Survey (YRBS) questionnaire; students are simply asked how many days they carried a weapon during the past 30 days. In the question that asks students about carrying a weapon at school, "on school property" was not defined for survey respondents.

SOURCE: Simone Robers et al., "Figure 14.1. Percentage of Students in Grades 9–12 Who Reported Carrying a Weapon at Least One Day during the Previous 30 Days, by Location and Sex: Various Years, 1993–2011," in *Indicators of School Crime and Safety: 2013*, National Center for Education Statistics and Bureau of Justice Statistics, June 2014, http://nces.ed.gov/pubs2014/2014042.pdf (accessed July 17, 2014)

Native students were the most likely demographic subgroup to carry a weapon anywhere, both in 1993, when 34.2% reported carrying a weapon, and in 2011, when 27.6% reported carrying a weapon. Among all demographic subgroups, weapon-carrying among African American students declined most significantly over this period, from 28.5% to 14.2%. The 2011 rates were 17% for white students, 16.2% for Hispanic students, 20.7% for Pacific Islander/Native Hawaiian students, and 23.7% for students of two or more races.

Weapon-carrying on school property varied less by race and ethnicity than did weapon-carrying anywhere. (See Table 7.4.) In 1993 Native American or Alaskan Native students (17.6%) had the highest prevalence of weapon-carrying on school property, followed by African American students (15%), Hispanic students (13.3%), and white students (10.9%). By 2011 both Native American or Alaskan Native students and students of two or more races had slightly higher prevalence rates of weapon-carrying on school property than other groups, at 7.5%, and there was little variation among white, African American, Hispanic, and Asian American students. A higher percentage (10.9%) of Pacific Islander/ Hawaiian students appeared to carry weapons on school property, but these data were less reliable. Figure 7.11 presents a visual interpretation of weapon-carrying among students by race, ethnicity, and location for 2011.

Table 7.5 provides information on weapon-carrying among high school students in 2013, distinguishing between the carrying of any weapon (such as a gun, knife, or club) and the carrying of a gun specifically. As with other statistics concerning weapons and violence, males were far more likely than females to carry either any weapon or a gun. Among females, there were not significant differences in the prevalence of either weapon- or gun-carrying by race and ethnicity, but there were significant differences by race and ethnicity among males. One-third (33.4%) of white male students reported carrying a weapon in the 30 days preceding the survey, and 10.7% reported carrying a gun. By comparison, 18.2% of African American male students reported carrying a weapon, and 9.8% reported carrying a gun; and 23.8% of Hispanic male students reported carrying a weapon, and 7.5% reported carrying a gun.

There was also considerable variation in weapon- and gun-carrying by state and urban location in 2013. As Table 7.6 shows, students in rural and southern states were considerably more likely to carry either weapons or guns than were students in coastal states or states with large urban populations. This pattern of variation applied to both males and females, so that states with high percentages of weapon- and gun-carrying males also tended to have high percentages of weapon- and gun-carrying females, but male rates were far higher in all states. The

TABLE 7.4

Percentage of students in grades 9–12 who reported carrying a weapon at least 1 day in the past 30 days, by location and selected student characteristics, selected years 1993–2011

Location and student characteristic	1993	1995	1997	1999	2001	2003	2005	2007	2009	2011
Anywhere (including on school property)[a]										
Total	22.1	20.0	18.3	17.3	17.4	17.1	18.5	18.0	17.5	16.6
Sex										
Male	34.3	31.1	27.7	28.6	29.3	26.9	29.8	28.5	27.1	25.9
Female	9.2	8.3	7.0	6.0	6.2	6.7	7.1	7.5	7.1	6.8
Race/ethnicity[b]										
White	20.6	18.9	17.0	16.4	17.9	16.7	18.7	18.2	18.6	17.0
Black	28.5	21.8	21.7	17.2	15.2	17.3	16.4	17.2	14.4	14.2
Hispanic	24.4	24.7	23.3	18.7	16.5	16.5	19.0	18.5	17.2	16.2
Asian[c]	—	—	—	13.0	10.6	11.6	7.0	7.8	8.4	9.1
Pacific Islander[c]	—	—	—	25.3	17.4	16.3	20.0	25.5	20.3	20.7
American Indian/Alaska Native	34.2	32.0	26.2	21.8	31.2	29.3	25.6	20.6	20.7	27.6
Two or more races[c]	—	—	—	22.2	25.2	29.8	26.7	19.0	17.9	23.7
Grade										
9th	25.5	22.6	22.6	17.6	19.8	18.0	19.9	20.1	18.0	17.3
10th	21.4	21.1	17.4	18.7	16.7	15.9	19.4	18.8	18.4	16.6
11th	21.5	20.3	18.2	16.1	16.8	18.2	17.1	16.7	16.2	16.2
12th	19.9	16.1	15.4	15.9	15.1	15.5	16.9	15.5	16.6	15:8
Urbanicity[d]										
Urban	—	—	18.7	15.8	15.3	17.0	—	—	—	—
Suburban	—	—	16.8	17.0	17.4	16.5	—	—	—	—
Rural	—	—	22.3	22.3	23.0	18.9	—	—	—	—
On school property[e]										
Total	11.8	9.8	8.5	6.9	6.4	6.1	6.5	5.9	5.6	5.4
Sex										
Male	17.9	14.3	12.5	11.0	10.2	8.9	10.2	9.0	8.0	8.2
Female	5.1	4.9	3.7	2.8	2.9	3.1	2.6	2.7	2.9	2.3
Race/ethnicity[b]										
White	10.9	9.0	7.8	6.4	6.1	5.5	6.1	5.3	5.6	5.1
Black	15.0	10.3	9.2	5.0	6.3	6.9	5.1	6.0	5.3	4.6
Hispanic	13.3	14.1	10.4	7.9	6.4	6.0	8.2	7.3	5.8	5.8
Asian[c]	—	—	—	6.5	7.2	6.6	2.8	4.1	3.6	4.3
Pacific Islander[c]	—	—	—	9.3	10.0	4.9	15.4	9.5	9.8	10.9
American Indian/Alaska Native	17.6	13.0	15.9	11.6	16.4	12.9	7.2	7.7	4.2	7.5
Two or more races[c]	—	—	—	11.4	13.2	13.3	11.9	5.0	5.8	7.5
Grade										
9th	12.6	10.7	10.2	7.2	6.7	5.3	6.4	6.0	4.9	4.8
10th	11.5	10.4	7.7	6.6	6.7	6.0	6.9	5.8	6.1	6.1
11th	11.9	10.2	9.4	7.0	6.1	6.6	5.9	5.5	5.2	4.7
12th	10.8	7.6	7.0	6.2	6.1	6.4	6.7	6.0	6.0	5.6
Urbanicity[d]										
Urban	—	—	7.0	7.2	6.0	5.6	—	—	—	—
Suburban	—	—	8.7	6.2	6.3	6.4	—	—	—	—
Rural	—	—	11.2	9.6	8.3	6.3	—	—	—	—

—Not available.

[a]The term "anywhere" is not used in the Youth Risk Behavior Survey (YRBS) questionnaire; students were simply asked how many days they carried a weapon during the past 30 days.

[b]Race categories exclude persons of Hispanic ethnicity.

[c]Before 1999, Asian students and Pacific Islander students were not categorized separately, and students were not given the option of choosing two or more races. Because the response categories changed in 1999, caution should be used in comparing data on race from 1993, 1995, and 1997 with data from later years.

[d]Refers to the Standard Metropolitan Statistical Area (MSA) status of the respondent's household as defined in 2000 by the U.S. Census Bureau. Categories include "central city of an MSA (Urban)," "in MSA but not in central city (Suburban)," and "not MSA (Rural)."

[e]In the question asking students about carrying a weapon at school, "on school property" was not defined for survey respondents.

Note: Respondents were asked about carrying "a weapon such as a gun, knife, or club."

SOURCE: Simone Robers et al., "Table 14.1. Percentage of Students in Grades 9–12 Who Reported Carrying a Weapon at Least 1 Day during the Previous 30 Days, by Location and Selected Student Characteristics: Selected Years, 1993 through 2011," in *Indicators of School Crime and Safety: 2013*, National Center for Education Statistics and Bureau of Justice Statistics, June 2014, http://nces.ed.gov/pubs2014/2014042.pdf (accessed July 17, 2014)

states in which male students were the most likely to carry weapons of any kind were Arkansas, where 42.2% of high school males reported carrying weapons, Wyoming (41.8%), Idaho (39.3%), Montana (38.5%), Alabama (36.1%), and Louisiana (36.1%). The states in which male students were the most likely to carry guns were Arkansas, where 20.7% reported gun-carrying, Louisiana (18.6%), Mississippi (18.5%), Wyoming (17.1%),

FIGURE 7.11

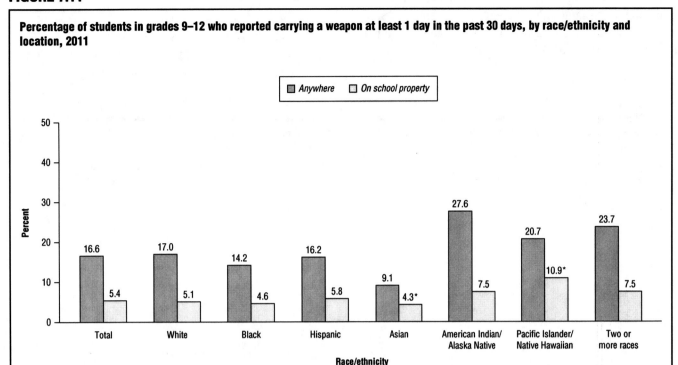

Percentage of students in grades 9–12 who reported carrying a weapon at least 1 day in the past 30 days, by race/ethnicity and location, 2011

*Interpret data with caution. The coefficient of variation (CV) for this estimate is between 30 and 50 percent.
Note: Race categories exclude persons of Hispanic ethnicity. The term "anywhere" is not used in the Youth Risk Behavior Survey (YRBS) questionnaire; students are simply asked how many days they carried a weapon during the past 30 days. In the question that asks students about carrying a weapon at school, "on school property" was not defined for survey respondents.

SOURCE: Simone Robers et al., "Figure 14.2. Percentage of Students in Grades 9–12 Who Reported Carrying a Weapon at Least One Day during the Previous 30 Days, by Race/Ethnicity and Location: 2011," in *Indicators of School Crime and Safety: 2013*, National Center for Education Statistics and Bureau of Justice Statistics, June 2014, http://nces.ed.gov/pubs2014/2014042.pdf (accessed July 17, 2014)

TABLE 7.5

Percentage of students in grades 9–12 who carried a weapon and who carried a gun, by sex, race/ethnicity, and grade, 2013

	Carried a weapon			Carried a gun		
	Female	Male	Total	Female	Male	Total
Category	%	%	%	%	%	%
Race/ethnicity						
White*	8.3	33.4	20.8	1.7	10.7	6.2
Black*	7.2	18.2	12.5	1.1	9.8	5.3
Hispanic	7.7	23.8	15.5	1.9	7.5	4.6
Grade						
9	8.6	26.4	17.5	1.9	9.1	5.5
10	9.2	26.4	17.8	1.6	8.4	5.0
11	5.9	30.5	17.9	1.1	10.5	5.7
12	7.5	29.5	18.3	1.6	9.9	5.7
Total	**7.9**	**28.1**	**17.9**	**1.6**	**9.4**	**5.5**

*Non-Hispanic.
Notes: Weapons include items such as, a gun, knife, or club. Incidents occurred on at least 1 day during the 30 days before the survey.

SOURCE: "Table 9. Percentage of High School Students Who Carried a Weapon and Who Carried a Gun, by Sex, Race/Ethnicity, and Grade—United States, Youth Risk Behavior Survey, 2013," in "Youth Risk Behavior Surveillance—United States, 2013," *Morbidity and Mortality Weekly Report*, vol. 63, no. 4, June 13, 2014, http://www.cdc.gov/mmwr/pdf/ss/ss6304.pdf (accessed July 17, 2014)

Montana (16.8%), and Alabama (14.9%). Students in large urban school districts were in general among the least likely to carry weapons or guns in 2013. The highest rates of weapon-carrying among male students in urban areas were found in the District of Columbia (26.9%), Duval County, Florida (26.9%), Baltimore, Maryland

(25%), and Houston, Texas (21.8%). The highest rates of gun-carrying among male students were found in Memphis, Tennessee (11.3%), Duval County, Florida (11.1%), Milwaukee, Wisconsin (10.6%), and Houston, Texas (10%).

Student Reports of Threats or Injuries

Table 7.7 shows trend data on weapons threats and injuries on school property between 1993 and 2011. The percentage of ninth-to-12th-grade students who were threatened or injured with a weapon, such as a gun, knife,

TABLE 7.6

Percentage of students in grades 9–12 who carried a weapon and who carried a gun, by sex, state, and urban area, 2013

	Carried a weapon			Carried a gun		
	Female	Male	Total	Female	Male	Total
Site	%	%	%	%	%	%
State surveys						
Alabama	9.7	36.1	23.1	3.2	14.9	9.2
Alaska	10.3	27.8	19.2	2.4	10.6	6.8
Arizona	9.8	24.7	17.5	2.5	7.8	5.2
Arkansas	11.4	42.2	27.1	5.2	20.7	13.3
Connecticut	—*	—	—	—	—	—
Delaware	5.5	23.3	14.4	1.6	8.8	5.2
Florida	7.4	23.8	15.7	—	—	—
Georgia	6.4	30.2	18.5	2.5	12.9	7.8
Hawaii	5.4	15.6	10.5	—	—	—
Idaho	14.2	39.3	27.0	—	—	—
Illinois	8.2	23.1	15.8	2.2	9.2	5.8
Kansas	5.9	25.8	16.1	—	—	—
Kentucky	7.6	33.5	20.7	2.5	12.3	7.5
Louisiana	9.8	36.1	22.8	4.8	18.6	11.9
Maine	—	—	—	—	—	—
Maryland	8.3	22.7	15.8	—	—	—
Massachusetts	4.8	18.1	11.6	0.6	4.9	2.9
Michigan	6.3	24.6	15.5	2.0	7.8	4.9
Mississippi	9.9	28.9	19.1	5.0	18.5	11.6
Missouri	10.6	33.2	22.2	—	—	—
Montana	12.4	38.5	25.7	3.7	16.8	10.5
Nebraska	—	—	—	—	—	—
Nevada	9.6	22.0	16.0	2.2	7.2	4.8
New Hampshire	—	—	—	—	—	—
New Jersey	3.9	16.6	10.2	0.8	5.0	2.9
New Mexico	12.3	31.9	22.2	3.5	11.2	7.4
New York	5.9	19.5	12.8	1.9	7.8	4.9
North Carolina	8.5	32.1	20.6	—	—	—
North Dakota	—	—	—	—	—	—
Ohio	6.5	21.5	14.2	—	—	—
Oklahoma	7.8	31.6	19.9	1.4	10.5	6.0
Rhode Island	—	—	—	2.2	8.6	5.6
South Carolina	11.1	30.9	21.2	3.0	12.8	8.1
South Dakota	—	—	—	—	—	—
Tennessee	8.2	30.2	19.2	2.2	11.6	7.0
Texas	9.4	27.2	18.4	1.9	9.9	6.0
Utah	7.6	26.6	17.2	1.9	10.0	6.1
Vermont	—	—	—	—	—	—
Virginia	7.3	23.7	15.8	3.3	11.0	7.3
West Virginia	8.7	38.9	24.3	2.7	13.3	8.2
Wisconsin	4.5	23.8	14.4	—	—	—
Wyoming	15.1	41.8	28.8	4.7	17.1	11.1
Median	8.2	27.5	18.4	2.4	10.8	6.9
Range	(3.9–15.1)	(15.6–42.2)	(10.2–28.8)	(0.6–5.2)	(4.9–20.7)	(2.9–13.3)
Large urban school district surveys						
Baltimore, MD	12.6	25.0	19.4	1.4	6.3	4.4
Boston, MA	8.4	15.6	12.1	1.6	4.8	3.3
Broward County, FL	6.0	14.0	10.2	1.4	2.9	2.3
Charlotte-Mecklenburg, NC	6.7	20.5	13.4	1.3	9.1	5.2
Chicago, IL	10.4	20.4	15.4	3.3	9.6	6.6
Detroit, MI	9.8	14.2	12.0	2.1	5.2	3.7
District of Columbia	13.1	26.9	20.0	—	—	—
Duval County, FL	11.6	26.9	19.0	3.3	11.1	7.2
Houston, TX	9.2	21.8	15.7	2.6	10.0	6.6
Los Angeles, CA	4.9	12.8	9.0	0.8	4.6	2.9
Memphis, TN	6.1	18.9	12.5	1.4	11.3	6.3
Miami-Dade County, FL	6.1	13.6	9.9	2.1	7.5	4.9
Milwaukee, WI	7.7	16.6	12.1	2.1	10.6	6.4
New York City, NY	5.1	11.2	8.3	1.2	3.8	2.5
Orange County, FL	7.2	16.9	12.3	1.8	5.7	4.0
Palm Beach County, FL	8.2	20.5	14.8	3.1	8.5	6.0

TABLE 7.6

Percentage of students in grades 9–12 who carried a weapon and who carried a gun, by sex, state, and urban area, 2013 [CONTINUED]

Site	Carried a weapon			Carried a gun		
	Female	Male	Total	Female	Male	Total
	%	%	%	%	%	%
Large urban school district surveys						
Philadelphia, PA	8.9	15.7	12.3	1.7	7.2	4.5
San Bernardino, CA	8.6	20.3	14.5	2.0	4.9	3.5
San Diego, CA	3.4	17.8	10.9	0.4	4.2	2.4
San Francisco, CA	5.4	12.7	9.2	0.7	4.2	2.6
Seattle, WA	—	—	—	2.9	8.4	6.0
Median	7.9	17.3	12.3	1.7	6.7	4.4
Range	(3.4–13.1)	(11.2–26.9)	(8.3–20.0)	(0.4–3.3)	(2.9–11.3)	(2.3–7.2)

*Non-Hispanic.

Notes: Weapons include items such as, a gun, knife, or club. Incidents occurred on at least 1 day during the 30 days before the survey.

SOURCE: "Table 10. Percentage of High School Students Who Carried a Weapon and Who Carried a Gun, by Sex—Selected U.S. Sites, Youth Risk Behavior Survey, 2013," in "Youth Risk Behavior Surveillance—United States, 2013," *Morbidity and Mortality Weekly Report*, vol. 63, no. 4, June 13, 2014, http://www.cdc.gov/mmwr/pdf/ss/ss6304.pdf (accessed July 17, 2014)

TABLE 7.7

Percentage of students in grades 9–12 who reported being threatened or injured with a weapon on school property during the last 12 months, by selected student characteristics and number of times threatened or injured, selected years 1993–2011

Number of times and year	Total	Sex		Race/ethnicity[a]							Grade			
		Male	Female	White	Black	Hispanic	Asian[b]	American Indian/ Alaska Native	Pacific Islander[b]	Two or more races[b]	9th grade	10th grade	11th grade	12th grade
At least once														
1993	7.3	9.2	5.4	6.3	11.2	8.6	—	11.7	—	—	9.4	7.3	7.3	5.5
1995	8.4	10.9	5.8	7.0	11.0	12.4	—	11.4!	—	—	9.6	9.6	7.7	6.7
1997	7.4	10.2	4.0	6.2	9.9	9.0	—	12.5!	—	—	10.1	7.9	5.9	5.8
1999	7.7	9.5	5.8	6.6	7.6	9.8	7.7	13.2!	15.6	9.3	10.5	8.2	6.1	5.1
2001	8.9	11.5	6.5	8.5	9.3	8.9	11.3	15.2!	24.8	10.3	12.7	9.1	6.9	5.3
2003	9.2	11.6	6.5	7.8	10.9	9.4	11.5	22.1	16.3	18.7	12.1	9.2	7.3	6.3
2005	7.9	9.7	6.1	7.2	8.1	9.8	4.6	9.8	14.5!	10.7	10.5	8.8	5.5	5.8
2007	7.8	10.2	5.4	6.9	9.7	8.7	7.6!	5.9	8.1!	13.3	9.2	8.4	6.8	6.3
2009	7.7	9.6	5.5	6.4	9.4	9.1	5.5	16.5	12.5	9.2	8.7	8.4	7.9	5.2
2011	7.4	9.5	5.2	6.1	8.9	9.2	7.0	8.2	11.3	9.9	8.3	7.7	7.3	5.9
Number of times, 2011														
0 times	92.6	90.5	94.8	93.9	91.1	90.8	93.0	91.8	88.7	90.1	91.7	92.3	92.7	94.1
1 time	3.1	3.6	2.5	2.9	3.6	3.1	2.7!	3.7	2.5!	3.9	4.1	2.5	3.0	2.5
2 or 3 times	1.9	2.6	1.3	1.5	2.3	2.8	2.0!	2.8	c	2.3	2.1	2.5	1.8	1.3
4 to 11 times	1.4	1.8	1.0	1.1	1.9	1.9	c	c	c	0.8!	1.2	1.7	1.3	1.2
12 or more times	1.0	1.5	0.4	0.7	1.1	1.4	c	c	c	2.8!	0.9	0.9	1.2	0.8

—Not available.

[a]Race categories exclude persons of Hispanic ethnicity.

[b]Before 1999, Asian students and Pacific Islander students were not categorized separately, and students were not given the option of choosing two or more races. Because the response categories changed in 1999, caution should be used in comparing data on race from 1993, 1995, and 1997 with data from later years.

[c]Reporting standards not met. Either there are too few cases for a reliable estimate or the coefficient of variation (CV) is 50 percent or greater.

!Interpret data with caution. The coefficient of variation (CV) for this estimate is between 30 and 50 percent.

Note: "On school property" was not defined for survey respondents. "Weapon" was defined as a gun, knife, or club for survey respondents. Detail may not sum to totals because of rounding.

SOURCE: Simone Robers et al., "Table 4.1. Percentage of Students in Grades 9–12 Who Reported Being Threatened or Injured with a Weapon on School Property during the Previous 12 Months, by Selected Student Characteristics and Number of Times Threatened or Injured: Selected Years 1993 through 2011," in *Indicators of School Crime and Safety: 2013*, National Center for Education Statistics and Bureau of Justice Statistics, June 2014, http://nces.ed.gov/pubs2014/2014042.pdf (accessed July 17, 2014)

---

or club, on school property ranged from 7.3% in 1993 to 9.2% in 2003. As happened with other crime indicators, threat reports increased from 1993 (7.3%) to 1995 (8.4%). Threat reports dropped through 1997 (7.4%), rose again through 2003 (9.2%), and then decreased substantially from 2003 to 2011 (7.4%). Males were

more likely than females to be threatened or injured with a weapon. In 2011, for example, 9.5% of male students were threatened or injured with a weapon, compared with 5.2% of female students who had similar experiences. In all the survey years, students in lower grades were more likely to be threatened than students in higher grades. In 2011, 11.3% of Pacific Islander or Hawaiian students, 9.9% of mixed-race students, 9.2% of Hispanic students, 8.9% of African American students, 8.2% of Native American or Alaskan Native students, 7% of Asian American students, and 6.1% of white students reported being threatened or injured with a weapon on school property.

The percentage of students who were threatened or injured with a weapon at school varied significantly by location. (See Table 7.8.) In 2011, 11.7% of Georgia students and 10.4% of Arizona students reported a threat or injury on school property, more than double the percentage of students in Wisconsin (5.1%) who reported a threat or injury. South Carolina students (9.2%), North Carolina students (9.1%), and District of Columbia students (8.7%) were among the other states with the highest prevalence of reported threats or injuries. Percentages for individual states varied from year to year, however, so there were no clear geographical judgments to be drawn based on these data.

Barring Guns from Schools

The Gun-Free Schools Act of 1994 required states to pass laws forcing schools that receive funding under the Elementary and Secondary Education Act of 1965 to expel for at least one year any student who brings a firearm to school. The U.S. Department of Education provides detail on the implementation of the act in periodic reports, the most recent of which, as of November 2014, was *Report on the Implementation of the Gun-Free Schools Act of 1994 in the States and Outlying Areas: School Years 2005–06 and 2006–07* (September 2010, http://www2.ed.gov/about/reports/annual/gfsa/gfsarp100610.pdf). The Department of Education indicates that 2,695 students were expelled from school during the 2006–07 academic year for carrying a firearm to school, a rate of 5.5 per 100,000 students. More than half (53%) of these students were expelled for carrying handguns, 10% were expelled for carrying rifles or shotguns to school, and the remaining 37% were expelled for carrying other types of firearms or destructive devices, such as starter pistols, bombs, and grenades.

The Department of Education also notes that the number of students expelled for carrying a firearm to school dropped by 22%, from 3,477 during the 1998–99 academic year to 2,695 during the 2006–07 academic year. Some states experienced reductions in the number of students expelled for carrying a firearm to school between the 2005–06 and the 2006–07 academic years. The states experiencing the greatest percentage of decreases in expulsions during this period were Delaware, Iowa, Kansas, New Hampshire, New Jersey, Rhode Island, and Wisconsin. The states that experi-

enced the greatest percentage of increases in expulsions during this period were Idaho, Indiana, Louisiana, and Minnesota.

TABLE 7.8

Percentage of students in grades 9–12 who reported being threatened or injured with a weapon on school property during the last 12 months, by state, selected years 2003–11

State	2003	2005	2007	2009	2011
United States*	9.2	7.9	7.8	7.7	7.4
Alabama	7.2	10.6	—	10.4	7.6
Alaska	8.1	—	7.7	7.3	5.6
Arizona	9.7	10.7	11.2	9.3	10.4
Arkansas	—	9.6	9.1	11.9	6.3
California	—	—	—	—	—
Colorado	—	7.6	—	8.0	6.7
Connecticut	—	9.1	7.7	7.0	6.8
Delaware	7.7	6.2	5.6	7.8	6.4
District of Columbia	12.7	12.1	11.3	—	8.7
Florida	8.4	7.9	8.6	8.2	7.2
Georgia	8.2	8.3	8.1	8.2	11.7
Hawaii	—	6.8	6.4	7.7	6.3
Idaho	9.4	8.3	10.2	7.9	7.3
Illinois	—	—	7.8	8.8	7.6
Indiana	6.7	8.8	9.6	6.5	6.8
Iowa	—	7.8	7.1	—	6.3
Kansas	—	7.4	8.6	6.2	5.6
Kentucky	5.2	8.0	8.3	7.9	7.4
Louisiana	—	—	—	9.5	8.7
Maine	8.5	7.1	6.8	7.7	6.8
Maryland	—	11.7	9.6	9.1	8.4
Massachusetts	6.3	5.4	5.3	7.0	6.8
Michigan	9.7	8.6	8.1	9.4	6.8
Minnesota	—	—	—	—	—
Mississippi	6.6	—	8.3	8.0	7.5
Missouri	7.5	9.1	9.3	7.8	—
Montana	7.1	8.0	7.0	7.4	7.5
Nebraska	8.8	9.7	—	—	6.4
Nevada	6.0	8.1	7.8	10.7	—
New Hampshire	7.5	8.6	7.3	—	—
New Jersey	—	8.0	—	6.6	5.7
New Mexico	—	10.4	10.1	—	—
New York	7.2	7.2	7.3	7.5	7.3
North Carolina	7.2	7.9	6.6	6.8	9.1
North Dakota	5.9	6.6	5.2	—	—
Ohio*	7.7	8.2	8.3	—	—
Oklahoma	7.4	6.0	7.0	5.8	5.7
Oregon	—	—	—	—	—
Pennsylvania	—	—	—	5.6	—
Rhode Island	8.2	8.7	8.3	6.5	—
South Carolina	—	10.1	9.8	8.8	9.2
South Dakota*	6.5	8.1	5.9	6.8	6.1
Tennessee	8.4	7.4	7.3	7.0	5.8
Texas	—	9.3	8.7	7.2	6.8
Utah	7.3	9.8	11.4	7.7	7.0
Vermont	7.3	6.3	6.2	6.0	5.5
Virginia	—	—	—	—	7.0
Washington	—	—	—	—	—
West Virginia	8.5	8.0	9.7	9.2	6.6
Wisconsin	5.5	7.6	5.6	6.7	5.1
Wyoming	9.7	7.8	8.3	9.4	7.3

—Not available.

*Data include both public and private schools.

Note: "On school property" was not defined for survey respondents. "Weapon" was defined as a gun, knife, or club for survey respondents. State-level data include public schools only, with the exception of data for Ohio and South Dakota. Data for the United States total, Ohio, and South Dakota include both public and private schools.

SOURCE: Simone Robers et al., "Table 4.2. Percentage of Public School Students in Grades 9–12 Who Reported Being Threatened or Injured with a Weapon on School Property at Least One Time during the Previous 12 Months, by State: Selected Years, 2003 through 2011," in *Indicators of School Crime and Safety: 2013*, National Center for Education Statistics and Bureau of Justice Statistics, June 2014, http://nces.ed.gov/pubs2014/2014042.pdf (accessed July 17, 2014)

SCHOOL SHOOTINGS

Mass shootings at U.S. schools became commonplace during the 1990s, shocking Americans across the political spectrum and leading to an evolving, ongoing debate about how best to prevent such atrocities in the future. The Columbine High School shooting in Littleton, Colorado, in 1999 marked a turning point in the public consciousness of such crimes and seemed destined to lead to a more concerted effort to prevent further mass shootings. However, the rate of school shootings increased during the early 2000s, and in recent years the country has witnessed both the deadliest school shooting in history—the Virginia Polytechnic Institute and State University killings in 2007—and arguably the most horrifying—the mass murder of first graders in Newtown, Connecticut, in 2012.

The following sections describe some of the better-known shooting incidents at U.S. high schools and universities in which multiple young people were shot. Although the stories and data related to school shootings in the United States are central to the current gun control debate, it should be noted that homicides at schools are exceedingly rare. Simone Robers et al. note in *Indicators of School Crime and Safety: 2013* (June 2014, http://nces.ed.gov/pubs2014/2014042.pdf) that in each year since the early 1990s, "youth homicides occurring at school remained at less than 2 percent of the total number of youth homicides." Thus, although high-profile mass murders in schools give many people the impression that American children are extremely vulnerable to violence, in fact the nation's schools are among the safest places children can be.

SECONDARY SCHOOL SHOOTINGS
Thurston High School, Springfield, Oregon

MAY 21, 1998. Fifteen-year-old Kip Kinkel (1982–) walked into the crowded cafeteria at Thurston High School in Springfield, Oregon, and opened fire with a semiautomatic rifle. The students Mikael Nickolauson and Ben Walker were killed, and 22 of their classmates were injured. Kinkel's parents were later found shot to death at their home. The year before, Kinkel's father had bought his son a Ruger .22-caliber semiautomatic rifle under the condition that he would use it only with adult supervision.

On September 24, 1999, as part of a plea agreement, Kinkel pleaded guilty to four counts of murder and 26 counts of attempted murder. On November 2, 1999, after a six-day sentencing hearing that included victims' statements and the testimony of psychiatrists and psychologists, Kinkel, by then aged 17, was sentenced to 111 years in prison without the possibility of parole.

Prompted by growing concerns over a rock-throwing incident that Kinkel had participated in and other behavioral problems, Faith Kinkel had taken her son to see a psychologist in January 1997, just over a year before the shooting. In this meeting the psychologist concluded that Kinkel was depressed, had difficulty managing his anger, and had shown a pattern of acting out his anger.

Columbine High School, Littleton, Colorado

APRIL 20, 1999. At 11:10 a.m., 18-year-old Eric Harris (1981–1999) arrived alone in the student parking lot at Columbine High School in Littleton, Colorado. Dylan Klebold (1982–1999), his 17-year-old classmate, arrived a short time later. Together, they walked to the school cafeteria carrying two large duffel bags, each concealing a 20-pound propane bomb set to detonate at exactly 11:17 a.m. After placing the duffel bags inconspicuously among hundreds of other backpacks and bags, Harris and Klebold returned to the parking lot to wait for the bombs to explode. As they waited, pipe bombs they had planted 3 miles (4.8 km) southwest of the high school exploded, resulting in a grass fire that was intended to divert the resources of the Littleton Fire Department and Jefferson County Sheriff's Office.

When their planted bombs failed to explode in the cafeteria, Harris and Klebold reentered the high school, this time via the west exterior steps, the highest point on campus with a view of the student parking lots and the cafeteria's entrances and exits. Both were wearing black trench coats that concealed 9mm semiautomatic weapons. They pulled out shotguns from a duffel bag and opened fire toward the west doors of the school, killing 17-year-old Rachel Scott. After entering the school, they killed 12 other victims, including a teacher, before finally killing themselves. Twenty-three more people were injured.

Within days, authorities had learned that three of the guns used in the massacre were purchased the year before by Klebold's girlfriend shortly after her 18th birthday. On May 3, felony charges were filed against 22-year-old Mark E. Manes for admittedly selling to Harris the TEC-DC9 semiautomatic handgun that he used in the shooting. On August 18, Manes pleaded guilty to the charge. The facts of this case as outlined came from *The Columbine High School Shootings: Jefferson County Sheriff Department's Investigation Report* (May 15, 2000).

THE COLUMBINE SHOOTERS. More than a year before the Columbine shooting, Harris and Klebold were arrested for breaking into a vehicle. In April 1998 both were placed in a juvenile diversion program and required to pay fines, attend anger management classes, and perform community service. Harris and Klebold successfully completed the diversion program and were released from it in February 1999 with their juvenile records cleared.

During the spring of 1998 Harris began keeping a diary, which was later recovered by authorities. In it he wrote of his desire to kill. In the only entry for 1999, Harris wrote of his and Klebold's preparations for what would become the Columbine massacre, including a detailed accounting of the weapons and bombs they intended to use.

After the Columbine shooting, Klebold's father, Tom Klebold, reported to investigators that his son had never showed any fascination with guns. The Klebolds told authorities that their son had been accepted to the University of Arizona, where he planned to major in computer science. Investigators who interviewed Klebold's friends and teachers heard him described as a nice, normal teenager.

Harris and Klebold left behind three videotapes documenting their plans and philosophies. The third videotape contained eight sessions taped from early April 1999 to the morning of the Columbine shooting on April 20, and showed some of their weapons and bombs, as well as recordings they had made of each other rehearsing for the shooting.

Red Lake High School, Red Lake, Minnesota

MARCH 21, 2005. The shooting that occurred at Red Lake High School was the nation's worst since the 1999 Columbine shooting. Red Lake High School is located on a Native American reservation in northern Minnesota. Jeffrey Weise (1988–2005), a 17-year-old junior, killed nine people and wounded seven in his shooting spree, and then shot and killed himself. Weise began his rampage by killing his grandfather—a tribal police officer—and his grandfather's female friend at their home, using a .22-caliber pistol of unknown origin. Weise then drove his grandfather's police cruiser to Red Lake High School. At the school, Weise used his grandfather's police-issued handguns and shotgun to kill a security guard, a teacher, and five students.

Weise came from a troubled background. He had lost his father to suicide in 1997. His mother was seriously brain-damaged in a car accident in 1999 and lived in a nursing home. Weise was thought to have posted messages on a neo-Nazi website. He called himself an "Angel of Death" and a "NativeNazi." He was often ridiculed and bullied by other students for his odd behavior.

Seven days after the shooting Louis Jourdain, the son of the tribal chairman, was arrested and charged with conspiracy. It was believed that he helped plot Weise's actions. Jourdain was tried as a juvenile, and in January 2006 he received a sentence, which, because of his juvenile status, was not made public. In July 2006 families of those injured and killed settled a lawsuit with the school district for $1 million.

Millard South High School, Omaha, Nebraska

JANUARY 5, 2011. After being suspended for a trespassing incident in which he drove his car on the school athletic fields, Robert Butler Jr. (1993–2011), a senior at Millard South High School and the son of an Omaha, Nebraska, police detective, entered the school shortly before 1:00 p.m. and mortally wounded Assistant Principal Vicki Kaspar in what appeared to be a targeted shooting. Butler wounded Principal Curtis Case on his way out of Kaspar's office and was later found dead of an apparent self-inflicted gunshot. The .40-caliber Glock pistol used in the shooting was believed to be his father's service revolver. Shortly before the incident Butler posted on his Facebook: "ur gonna here about the evil sh** I did but that f***ing school drove me to this. I wont u guys to remember me for who I was b4 this ik. I greatly affected the lives of the families ruined but I'm sorry. goodbye."

Chardon High School, Chardon, Ohio

FEBRUARY 27, 2012. At about 7:30 a.m. on February 27, 2012, Thomas "T. J." Lane (1994–) opened fire with a .22-caliber semiautomatic Ruger handgun in the Chardon High School cafeteria as he and other students waited for bus transportation to other educational sites. Six students were hit by the gunfire; one died at the scene, and two other students died the following day from their wounds. As the incident happened, Lane was confronted by an unarmed football coach at the school and chased from the building. He was apprehended outside by law enforcement and detained in a juvenile facility as prosecutors sought to try him as an adult. At a hearing in June 2012 Lane registered a plea of not guilty by reason of insanity. He reportedly admitted to the shooting but told authorities that he did not know why he did it. The gun used in the case had been legally purchased by a relative and was allegedly stolen by Lane the night before the shooting. Lane later changed his plea, pleading guilty to the murders of his three classmates as well as to other charges, and in 2013 he was sentenced to life imprisonment without the possibility for parole.

POSTSECONDARY SCHOOL SHOOTINGS

Virginia Polytechnic Institute and State University, Blacksburg, Virginia

APRIL 16, 2007. The Virginia Polytechnic Institute and State University (Virginia Tech) massacre was the deadliest shooting rampage by a single gunman in U.S. history. A Virginia judge had declared Seung-Hui Cho (1984–2007) mentally ill. Nonetheless, this Virginia Tech student was able to purchase two handguns. On the morning of

April 16, 2007, Cho entered a residence hall on campus, where he shot and killed a female student and a male resident assistant. Hours later, after returning to his dorm room to change out of his bloodied clothes and to delete various files from his computer, Cho entered a classroom building on campus and began shooting students and professors. When he finished, Cho had killed 32 students and faculty members and had injured 15 others. Cho committed suicide by shooting himself in the head.

Northern Illinois University, DeKalb, Illinois

FEBRUARY 14, 2008. The Northern Illinois University's (NIU) Department of Safety details in *The Report of the February 14, 2008, Shootings at Northern Illinois University* (March 2010, http://www.niu.edu/feb14report/Feb14report.pdf) the story of what happened on that deadly day in Cole Hall, where gunfire killed five students and wounded 21 others, including a professor. The shooter was Steven Kazmierczak (1980–2008), a once-successful student at NIU, who shot and killed himself when his rampage from the stage of an auditorium filled with oceanography students was over. The former NIU student first fired six rounds from a shotgun and then another 50 rounds from a 9mm Glock semiautomatic pistol; 55 unused rounds were found at the scene. Kazmierczak had a history of mental health problems and had chosen to discontinue his medication prior to the February 14 shooting. Officials could determine no motive for the killings other than his mental illness and suggested that Kazmierczak chose NIU as the site simply because he was a former student there and was familiar with the campus.

Oikos University, Oakland, California

APRIL 2, 2012. Seven people were killed and three others were wounded when One L. Goh (1968–) opened fire at about 10:30 a.m. on April 2, 2012, with a .45-caliber handgun in a nursing classroom at Oikos University in Oakland, California. The Korean-born Goh was identified as a former student at the private school, which was founded in 2004 and is affiliated with the Praise God Korean Church of Oakland. Officials at the school were unable to identify a motive in the shooting. Goh was in financial difficulty and had dropped out of the school, which is located in a tight-knit Korean American community, but he had not been expelled. On the morning of the attack he attempted to meet with a school administrator, but when she was not in her office, he forced another worker at the school to accompany him into a classroom where he began shooting at the students. Goh left in a stolen vehicle and threw his gun, which he had purchased legally in California, into a nearby estuary before turning himself in to authorities that afternoon. In late 2013 a court-appointed psychiatrist found Goh mentally incompetent to stand trial, and

after a hearing to confirm this assessment, he was sent to the Napa State Hospital in Northern California, where he would be treated until considered fit to stand trial.

WHEN ADULTS SHOOT SCHOOLCHILDREN

Although most school shootings are carried out by students, school property occasionally becomes the site of gun violence perpetrated by adults against children. This was the case during the fall of 2006, when in just one week two schools were the scenes of violence involving adult males who entered school property intending to kill young female students. It was also the case in February 2010, when Littleton, Colorado, was once again the site of a school shooting, but this time it was an adult targeting middle school students. Just over two and a half years later, in December 2012, an elementary school in Newtown, Connecticut, was the site of a horrific attack that resulted in the deaths of 26 people, 20 of them children between the ages of six and seven; the shooter also died.

Platte Canyon High School, Bailey, Colorado

SEPTEMBER 27, 2006. Fifty-three-year-old Duane Morrison (1952–2006) entered a second-floor classroom at Platte Canyon High School in Bailey, Colorado, with two handguns and a backpack that he claimed contained a bomb. Morrison, who had no apparent ties to the school, took six female students hostage, ordering the rest of the students out of the room. The students who were allowed to leave told police that Morrison seemed to choose blonde girls of small stature to keep as hostages. Several hours later, after Morrison had released four of the girls, police burst through the classroom door. One of the girls escaped unharmed, but before Morrison turned his gun on himself, he shot and killed 16-year-old Emily Keyes. Police were unsure of Morrison's motive, but the girls who had been held hostage confirmed that Morrison—who was described as a "drifter" with a record of minor criminal offenses—had sexually assaulted them during the standoff.

West Nickel Mines Amish School, Nickel Mines, Pennsylvania

OCTOBER 2, 2006. Less than one week after the incident in Bailey, Colorado, a 32-year-old milk truck driver entered a one-room Amish schoolhouse in Lancaster County, Pennsylvania. Armed with three guns, a stun gun, two knives, 600 rounds of ammunition, and a number of instruments believed to be intended for torture and sexual assault, the gunman ordered all male students and adult females to leave the building. He then lined up the remaining 10 children—all girls, aged six to 13—against the blackboard and bound their feet. After escaping, a teacher at the school ran to a farmhouse that had a telephone and called police. When the police arrived, the gunman shot all 10 girls and then himself. Five of the girls died.

Later identified as Charles C. Roberts IV (1973–2006), the gunman lived with his wife and children in Nickel Mines. Family and friends, including many of his Amish neighbors, were shocked by Roberts's actions, maintaining that he had appeared to be a devoted husband and father. Before taking his own life, Roberts called his wife and explained that he was plagued by guilt for having molested two young relatives when he was 12 years old and that he had recently experienced recurring dreams of molesting more girls. Police, however, were unable to confirm Roberts's story.

Deer Creek Middle School, Littleton, Colorado

FEBRUARY 23, 2010. As buses were leaving Deer Creek Middle School at the end of the school day, several students heard two shots ring out. Within seconds, 32-year-old Bruco Strongeagle Eastwood (1977?–) was tackled by a seventh-grade math teacher as Eastwood tried to reload his high-powered rifle. Lying wounded were eighth graders Matt Thieu and Reagan Webber. Eastwood, a former student of Deer Creek Middle School, had previously visited the school and was seen inside the school shortly before the shooting. A student standing nearby the victims said that an adult approached and asked the students if they attended the school. When Thieu and Webber responded affirmatively, Eastwood shot them. Both students survived the shooting. Although police revealed no motive for the shooting, they did note that Eastwood was apparently depressed about never graduating from high school and being unable to obtain a general equivalency diploma.

Sandy Hook Elementary School, Newtown, Connecticut

DECEMBER 14, 2012. Twenty-year-old Adam Lanza (1992?–2012) shot his mother to death at their home in Newtown and then proceeded to Sandy Hook Elementary School, where he shot his way inside. Armed with a variety of weapons, including two handguns and an assault rifle, he killed 20 first-grade students and six adult staff members before killing himself. In the aftermath of the atrocity, details about Lanza's mental-health problems emerged, as did details of his intimate knowledge of handguns and assault weapons. His mother was a firearms collector and an avid target shooter, and she passed this enthusiasm onto her son.

The incident drew international media attention and prompted intense debate over the adequacy of the nation's gun control regulations. On January 16, 2013 (http://www.whitehouse.gov/the-press-office/2013/01/16/remarks-president-and-vice-president-gun-violence), President Barack Obama (1961–) presented a plan to reduce mass gun violence that included 23 executive actions and proposals for four major pieces of legislation. Vowing to spend political capital in what would undoubtedly be difficult negotiations with the opponents of policy reform, Obama said, "I will put everything I've got into this." Political opposition to increased gun control quickly scuttled reformers' hopes for change in the months after Newtown, and many gun rights advocates, in assessing the legacy of Newtown, emphasized the poor national mental-health infrastructure as a more important factor in the shooting than the ease of access to guns.

Post-Newtown School Safety Efforts

Although the Sandy Hook shooting's effects on gun control laws was minimal by comparison with the initial horror the event provoked, the tragedy did bring about substantial changes in school security standards nationwide. Brendan O'Brien and Ian Simpson report in "U.S. Schools Look to Guards, Technology a Year after Sandy Hook" (Reuters.com, December 11, 2013) that, in the year after the shooting, school administrators nationwide began spending unprecedented sums on security technologies such as electronically controlled doors, bulletproof glass or security screens for windows, better lighting, emergency communications systems, and surveillance cameras. U.S. schools were expected to spend $4.9 billion annually on such security hardware by 2017, up from the $2.7 billion spent in 2012. Also in response to Sandy Hook, the presence of school resource officers, as police officers assigned to public schools are called, became a norm toward which school districts aspired, rather than a sign of trouble. O'Brien and Simpson quote Ronald Stephens, the executive director of the nonprofit National School Safety Center, who said, "It used to be if you had a cop on the campus, people would see it as something wrong with the school. Now, it's seen as an advantage."

This approach to school security was the preferred outcome of many gun rights advocates, including Wayne LaPierre, the chief executive officer of the National Rifle Association of America, who responded to the massacre by calling for the placing of armed officers in all 99,000 public schools in the United States. Additionally, as O'Brien and Simpson report, seven states changed their laws in response to Newtown to allow school employees and even, in some cases, ordinary citizens, to carry firearms in public schools.

CHILDREN INJURED AND KILLED BY GUNFIRE

The Children's Defense Fund (CDF), a charitable organization that focuses on the needs of poor and minority children and those with disabilities, regularly reports on the number of children who lose their lives to gun violence. In *Protect Children, Not Guns 2013* (July 2013, http://www.childrensdefense.org/child-research-data-publications/data/protect-children-not-guns-2013.pdf), the CDF indicates that in 2010, 18,270 American children and teens died or were injured by guns. This represents one child death or injury per half hour, 50 child deaths or injuries per day, and 351 child deaths or injuries per week. As the CDF further notes, "More children and teens die from guns every three days than died in the Newtown massacre," and the total number of children and teens who died in 2010 (2,694) "would fill 134 classrooms of 20 children."

FIREARMS AND YOUTH SUICIDE

Although homicide rates have declined dramatically since the 1990s, suicide rates have declined only slightly since that time. In 1950 the age-adjusted rate of death by suicide for all ages was 13.2 per 100,000, and in 2010 it was 12.1 per 100,000, with slight decreases and increases from decade to decade in between. (See Table 6.6 in Chapter 6.) In contrast, the suicide rate for young people rose precipitously over this same period. The rate of death by suicide nearly tripled for youth aged 15 to 24 years between 1950 (4.5 per 100,000) and 1990 (13.2 per 100,000) and then declined slightly to 10.5 per 100,000 in 2010. Between 1950 and 2010, then, the youth suicide rate more than doubled. As Table 7.9 shows, suicide was the 10th-leading cause of death among all age groups in 2011, the fourth-leading cause of death among children aged five to 14 years, and the second-leading cause of death among young people aged 15 to 24 years. Among young people aged 15 to 24 years, there were 4,688 deaths by suicide in 2011.

This increase in the youth suicide rate was largely a result of increases among young males. The rate of death by suicide for males aged 15 to 24 years increased from 6.5 per 100,000 in 1950 to 16.9 per 100,000 in 2010, whereas the rate for females in the same age group increased from 2.6 per 100,000 in 1950 to 3.9 per 100,000 in 2010. (See Table 6.6 in Chapter 6.) According to the CDC, in "Suicide Prevention: Youth Suicide" (January 9, 2014, http://www.cdc.gov/violenceprevention/pub/youth_suicide.html), males account for 81% of suicide deaths among people aged 10 to 24 years, and females account for 19%.

The CDC notes that firearms account for 45% of suicides among young people aged 10 to 24 years, while suffocation accounts for 40% and poisoning for 8%. Firearms are different from other suicide methods in one key aspect: attempts are much more likely to be fatal. Other means of attempting suicide are less reliable, take time, and can be reversed, often through intervention from loved ones or through emergency medical care. In "The Epidemiology of Case Fatality Rates for Suicide in the Northeast" (*Annals of Emergency Medicine*, vol. 43, no. 6, June 2004), Matthew Miller, Deborah Azrael, and David Hemenway of the Harvard Injury Control Research Center (HICRC) find that the fatality rate of suicide attempts with a firearm exceeds 90%, compared with a fatality rate of less than 5% for other common suicide methods, such as drug overdoses and cutting/piercing.

The lethality of firearms is of particular concern in adolescence and young adulthood, a time of life when many people begin experiencing pronounced emotional and social difficulties without having the life experience or maturity to put them into perspective. The CDC indicates that approximately 157,000 youth are hospitalized for suicide attempts annually. This number, when compared with the 4,600 young people who die from suicide each year, suggests that adolescents and young adults are particularly likely to entertain the possibility of suicide without necessarily being resolute in their determination to die. However, young people who attempt suicide with firearms are almost certain to die even if their suicidal impulses are temporary rather than deeply rooted.

U.S. suicide rates are highest in rural areas and in the western states, a fact that many researchers have linked with the prevalence of gun ownership in such locations. The American Association of Suicidology, a membership organization that encompasses a range of professionals in the fields of suicide prevention and intervention, presents in "USA State Suicide Rates and Rankings among the Elderly and Young, 2011" (2014, http://www.suicidology.org/Portals/14/docs/Resources/FactSheets/2011AgeSpecific Data.pdf) suicide rates by state and age group, based on statistics from the CDC's Web-Based Injury Statistics Query and Reporting System database. The states with the highest youth suicide rates in 2011 were Alaska (33.2 per 100,000), South Dakota (22.5), Idaho (22.2), Wyoming (21.7), and Montana (20.7). There are no comprehensive statistics on the prevalence of gun ownership, given constraints on the collection of such data as well as the large number of guns that are inherited or purchased privately. However, the article "States with the Most Legal Guns in 2012" (DailyBeast.com, December 15, 2012) compares statistics from the Federal Bureau of Investigation's National Instant Criminal Background Check System to state population statistics to determine the states with the highest number of background checks (which correspond to legal gun purchases) per capita. Montana ranked second nationally in background checks per capita, Alaska ranked fifth, Wyoming ranked sixth, South Dakota ranked seventh, and Idaho ranked 12th.

This correlation between high rates of youth suicide and high rates of gun ownership has been substantiated by numerous scholarly studies, including many conducted by HICRC researchers and published between 1999 and 2007. The methodology and findings of these studies are summarized at the center's website (http://www.hsph.harvard.edu/hicrc/firearms-research/

gun-ownership-and-use). Renee M. Johnson et al. of the HICRC find in "Who Are the Owners of Firearms Used in Adolescent Suicides?" (*Suicide and Life Threatening* *Behavior*, vol. 40, no. 6, December 2010) that a majority of adolescents who commit suicide with firearms use weapons owned by their parents.

TABLE 7.9

Deaths and death rates for the 10 leading causes of death among young people, 2011

[Data are based on a continuous file of records received from the states. Rates are per 100,000 population in specified group. Figures for 2011 are based on weight and are rounded to the nearest individual, so categories may not add to toals or subtotals.]

Rank[a]	Cause of death (based on the International Classification of Diseases, Tenth Revision, 2008 Edition, 2009) and age	Number	Rate
	All ages[b]		
—	All causes	2,512,873	806.5
1	Diseases of heart	596,339	191.4
2	Malignant neoplasms	575,313	184.6
3	Chronic lower respiratory diseases	143,382	46.0
4	Cerebrovascular diseases	128,931	41.4
5	Accidents (unintentional injuries)[c]	122,777	39.4
—	Motor vehicle accidents	34,676	11.1
—	All other accidents	88,101	28.3
6	Alzheimer's disease[c]	84,691	27.2
7	Diabetes mellitus	73,282	23.5
8	Influenza and pneumonia[d]	53,667	17.2
9	Nephritis, nephrotic syndrome and nephrosis[e]	45,731	14.7
10	Intentional self-harm (suicide)	38,285	12.3
—	All other causes	650,475	208.8
	1–4 years		
—	All causes	4,214	26.1
1	Accidents (unintentional injuries)[c]	1,346	8.3
—	Motor vehicle accidents	416	2.6
—	All other accidents	930	5.8
2	Congenital malformations, deformations and chromosomal abnormalities[c]	483	3.0
3	Assault (homicide)	370	2.3
4	Malignant neoplasms	352	2.2
5	Diseases of heart	158	1.0
6	Influenza and pneumonia[d]	96	0.6
7	Septicemia	59	0.4
8	Chronic lower respiratory diseases	44	0.3
9	In situ neoplasms, benign neoplasms and neoplasms of uncertain or unknown behavior	43	0.3
9	Cerebrovascular diseases	43	0.3
—	All other causes	1,220	7.5
	5–14 years		
—	All causes	5,395	13.1
1	Accidents (unintentional injuries)[c]	1,613	3.9
—	Motor vehicle accidents	867	2.1
—	All other accidents	746	1.8
2	Malignant neoplasms[c]	865	2.1
3	Congenital malformations, deformations and chromosomal abnormalities	356	0.9
4	Intentional self-harm (suicide)	281	0.7
5	Assault (homicide)	269	0.7
6	Diseases of heart	185	0.5
7	Chronic lower respiratory diseases	134	0.3
8	Influenza and pneumonia[d]	112	0.3
9	Cerebrovascular diseases	83	0.2
10	In situ neoplasms, benign neoplasms and neoplasms of uncertain or unknown behavior	72	0.2
—	All other causes	1,425	3.5
	15–24 years		
—	All causes	29,605	67.6
1	Accidents (unintentional injuries)[c]	12,032	27.5
—	Motor vehicle accidents	6,984	15.9
—	All other accidents	5,048	11.5
2	Intentional self-harm (suicide)[c]	4,688	10.7
3	Assault (homicide)	4,508	10.3
4	Malignant neoplasms	1,609	3.7
5	Diseases of heart	948	2.2
6	Congenital malformations, deformations and chromosomal abnormalities	429	1.0
7	Influenza and pneumonia[d]	213	0.5
8	Cerebrovascular diseases	186	0.4
9	Pregnancy, childbirth and the puerperium	166	0.4
10	Chronic lower respiratory diseases	160	0.4
—	All other causes	4,666	10.7

TABLE 7.9

Deaths and death rates for the 10 leading causes of death among young people, 2011 [CONTINUED]

[Data are based on a continuous file of records received from the states. Rates are per 100,000 population in specified group. Figures for 2011 are based on weight and are rounded to the nearest individual, so categories may not add to toals or subtotals.]

—Category not applicable.
[a]Rank based on number of deaths.
[b]Includes deaths under age 1 year.
[c]New ICD-10 subcategories were introduced for the existing (victim of earthquake).
[d]New ICD-10 code (Human metapneumovirus pneumonia) was added to the category in 2011.
[e]New subcategories replaced previous ones for chronic kidney disease in 2011. Changes affect comparability with previous year's data.
Notes: For certain causes of death such as unintentional injuries, homicides, suicides, and respiratory diseases, preliminary and final data differ because of the truncated nature of the preliminary file.
Data are subject to sampling or random variation.

SOURCE: Adapted from Donna L. Hoyert and Jiaquan Xu, "Table 7. Deaths and Death Rates for the 10 Leading Causes of Death in Specified Age Groups: United States, Preliminary 2011," in "Deaths: Preliminary Data for 2011," *National Vital Statistics Reports*, vol. 61, no. 6, October 10, 2012, http://www.cdc.gov/nchs/data/nvsr/nvsr61/nvsr61_06.pdf (accessed August 18, 2014)

CHAPTER 8
PUBLIC ATTITUDES TOWARD GUN CONTROL

A DEEPLY PERSONAL ISSUE

During the past 50 years, gun control has been a prominent issue, fueled by firearm-related political assassinations, assassination attempts, and violent crimes that became top news stories. Americans mourned the deaths of President John F. Kennedy (1917–1963); his brother Senator Robert F. Kennedy (1925–1968; D-NY); and the civil rights leader Martin Luther King Jr. (1929–1968). Presidents Gerald R. Ford (1913–2006) and Ronald Reagan (1911–2004) were victimized by would-be assassins, as were the presidential candidate George C. Wallace (1919–1998), the civil rights leader Vernon E. Jordan (1935–), and Representative Gabrielle Giffords (1970–; D-AZ), among many others.

Debates over the easy availability of guns in the United States also erupt in the wake of school shootings, which have increased in frequency since the 1990s. The most notorious of these incidents include the 1999 Columbine High School massacre in Littleton, Colorado, the 2007 Virginia Polytechnic Institute and State University killings, and the 2012 Sandy Hook Elementary School shooting in Newtown, Connecticut, each of which led to calls for tightened gun control regulations. (For a summary of some of the most high-profile mass shootings since 1949, see Table 5.7 in Chapter 5.)

Political assassinations and school shootings attract more media attention than other types of firearm violence, but they are, in fact, extremely rare, representing only a small fraction of the overall deaths and injuries caused by gun use annually. It is the use of guns in everyday crime and suicide that most troubles public health professionals, if not ordinary Americans. As Table 6.7 in Chapter 6 shows, 300,764 people were killed by firearms in the United States between 2002 and 2011. An increasing number of these deaths are suicides, and although many of those who commit suicide with firearms would likely commit suicide with or without guns,

guns are indisputably the most deadly means of doing so. This is of particular concern among adolescents and young adults, who frequently mistake temporary problems for permanent ones, and whose suicide rate has risen dramatically since 1950. (See Chapter 7.)

However, as many gun rights advocates point out, it is difficult to prove that tighter gun ownership and purchase requirements would prevent many of these crimes. Criminals often obtain weapons illegally, whereas law-abiding citizens are the primary purchasers of guns in stores that conduct background checks. Many people who use guns to commit suicide, including most young people, obtain those guns from their own home, following legal purchases by parents or other household members. Although gun control advocates point to numerous studies showing that the presence of guns in the home positively correlates with higher rates of homicide and suicide, it is unclear that placing more restrictions on the purchase of firearms would address such deaths. Meanwhile, as described in Chapter 1, the right to keep and bear arms is interwoven with the history and political foundations of the United States in such a way that many gun owners see their firearm possession as one of the basic features of their identity as an American. Thus, the gun control debate touches on some of the most emotionally charged territory in U.S. politics.

Some Americans are convinced that more federal regulation of firearms is necessary to reduce the number of murders and injuries that are inflicted with guns and to ensure a safer, more civilized society. They see firearms as the cause of a colossal and unconscionable waste of life and health and note that countries that restrict firearm ownership have much lower rates of firearm death. Their opponents on the other side of the debate insist that the right to bear arms is guaranteed by long-standing custom and by the Second Amendment to the U.S. Constitution. These gun rights advocates believe that no cyclical increase in crime, no mass killing, or any rash of political murders should lead the nation to violate the Constitution and the individual rights it guarantees.

Both supporters and opponents of gun control agree that some means should be found to keep guns out of the hands of criminals. Not surprisingly, the two sides approach this issue differently. The two different strategies for gun control involve deterrence (discouraging by instilling fear) and interdiction (legally forbidding the use of). Advocates of deterrence, most notably the Second Amendment Foundation and the National Rifle Association of America (NRA), recommend consistent enforcement of current laws and instituting tougher penalties to discourage individuals from using firearms in crimes. They maintain that interdiction will not have any effect on crime but will strip away the constitutional rights and privileges of law-abiding Americans by taking away their right to own guns.

Advocates of interdiction, led by organizations such as the Brady Campaign to Prevent Gun Violence, the Law Center to Prevent Gun Violence, and the Violence Policy Center, believe controlling citizens' access to firearms will reduce crime. Therefore, they favor restrictions on public gun ownership.

EVALUATING PUBLIC OPINION POLLS

Public opinion polls, like all sources of information, must be used with care. Pollsters select sample populations because it is impossible to interview every American on a given question. The selection is usually performed randomly using computers. Most major pollsters interview between 1,000 and 2,000 people to establish a valid sample. Other pollsters may interview far fewer than 500, and the sample may be too small to fairly represent the opinions of all adult Americans. Generally, the larger the sample, the greater the chance for an adequate representation and a valid result.

The polling errors that concern most people are those caused by bias in the presentation of questions, which may influence the response. For example, is the question vague? Is it too long? Is it threatening? Is it leading? If the questions are asked in an in-person interview, was the interviewer too forceful or threatening? Did respondents provide answers they thought would please interviewers? Were respondents disqualified because of membership in gun control or gun rights organizations? What was the purpose of the poll? Who hired the polling organization, and what is its stance on the issue?

Respondents may be unwilling to candidly discuss their use of weapons. In addition, polling does not always determine how important a person considers an issue to be. The issue may be of absolutely no concern to the respondent, but when asked, the respondent then thinks

about the topic and provides an answer. Five minutes after the question has been asked, the issue may completely disappear from his or her mind.

The polling organization might not include the number of times there was no response. If these "no replies" come predominantly from one group, they might influence the poll so that it does not truly represent the national opinion on a given issue. Pollsters are aware of these weaknesses, so they usually indicate how reliable they consider their polls to be. As a result, many polls indicate an accuracy rate of plus or minus (+/−) 2% to 4%.

PUBLIC OPINION ON GUN CONTROL
The Demographics of Gun Ownership

Gun ownership rates vary significantly by region, state, location, sex, race and ethnicity, and other demographic factors. Rich Morin explains in "The Demographics and Politics of Gun-Owning Households" (July 15, 2014, http://www.pewresearch.org/fact-tank/2014/07/15/the-demographics-and-politics-of-gun-owning-house holds) that the Pew Research Center, a nonpartisan research and polling firm, established the prevalence of gun ownership by demographic and political characteristics as part of its American Trends Panel survey conducted in April and May 2014.

By region, Pew found that the percentage of gun-owning households among all households is similar in the West (34%), the Midwest (35%), and the South (38%). (See Figure 8.1.) In contrast, the percentage of gun-owning households in the Northeast (27%) is much lower. Accordingly, residents of northeastern states, on average, favor stricter gun control laws than residents in other regions, with the notable exceptions of California, Hawaii, and Illinois, which have comparably strict gun control laws at the state level. As Figure 3.4 in Chapter 3 shows, seven of the 10 states that were given an A or B rating (on an A to F scale) for the effectiveness of their gun control laws in 2013 by the Brady Campaign to Prevent Gun Violence were in the Northeast.

As part of its mid-2014 survey, Pew also assessed gun ownership rates by gender, race and ethnicity, age, location, and the presence of children in the home. It found a significant disparity by gender: 38% of American men reported the presence of a gun in their home, compared with 31% of women. (See Figure 8.2.) It also found that white Americans are much more likely to own guns than are African Americans and Hispanics. Whereas 41% of non-Hispanic white Americans reported the presence of a gun in their home, only 19% of African Americans and 20% of Hispanics reported a gun in the home. Gun ownership rates also increase with age. The presence of a gun in the home was reported by 26% of 18- to 29-year-olds, 32% of 30- to 49-year-olds, 40% of 50- to 64-year-olds, and 40% of those aged 65 years and older. More

FIGURE 8.1

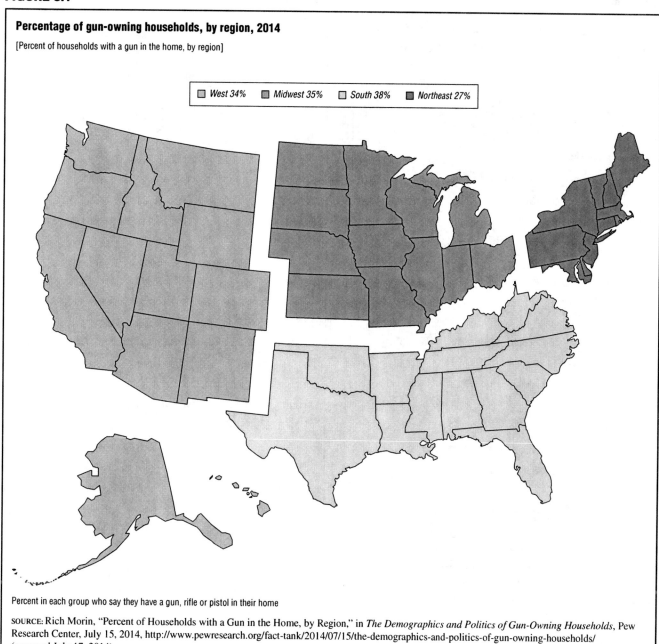

Percentage of gun-owning households, by region, 2014

[Percent of households with a gun in the home, by region]

West 34%   Midwest 35%   South 38%   Northeast 27%

Percent in each group who say they have a gun, rifle or pistol in their home

SOURCE: Rich Morin, "Percent of Households with a Gun in the Home, by Region," in *The Demographics and Politics of Gun-Owning Households*, Pew Research Center, July 15, 2014, http://www.pewresearch.org/fact-tank/2014/07/15/the-demographics-and-politics-of-gun-owning-households/ (accessed July 17, 2014)

than half (51%) of rural Americans reported the presence of a gun in the home in 2014, compared with 36% of suburban residents and 25% of urban residents. Gun ownership rates were not significantly different for households that included children under the age of 18 years (35%) than for the population at large (34%), but there was a nontrivial difference in the proportion of households with children aged zero to four years (33%) and children aged five to 11 years (33%) who reported guns in the home versus the proportion of households with children aged 12 to 17 years (37%).

Gun ownership rates also track the division between the two major political parties in the United States, as well as divisions among broader ideological categories. Whereas almost half (49%) of Republicans reported the presence of guns in their home in 2014, less than a quarter (22%) of Democrats did. (See Figure 8.2.) Independents (37%) were slightly closer to Republicans than to Democrats in their gun ownership status. Among those Americans who identify as conservative, 41% reported guns in the home, compared with 36% who identify as moderate and 23% who identify as liberal.

Although some gun owners approve of gun control regulations and some people with no interest in owning guns approve of others' freedom to own guns without substantial restrictions, differences of opinion nevertheless

FIGURE 8.2

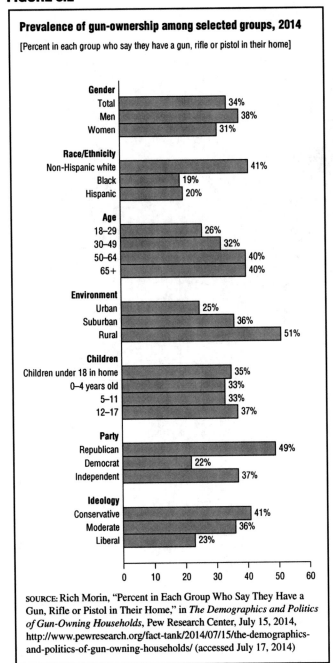

Prevalence of gun-ownership among selected groups, 2014

[Percent in each group who say they have a gun, rifle or pistol in their home]

SOURCE: Rich Morin, "Percent in Each Group Who Say They Have a Gun, Rifle or Pistol in Their Home," in *The Demographics and Politics of Gun-Owning Households*, Pew Research Center, July 15, 2014, http://www.pewresearch.org/fact-tank/2014/07/15/the-demographics-and-politics-of-gun-owning-households/ (accessed July 17, 2014)

FIGURE 8.3

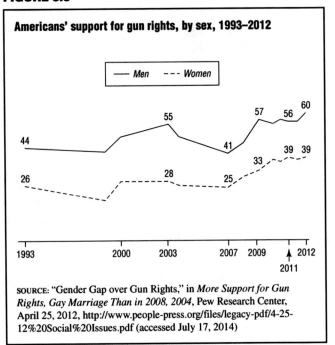

Americans' support for gun rights, by sex, 1993–2012

SOURCE: "Gender Gap over Gun Rights," in *More Support for Gun Rights, Gay Marriage Than in 2008, 2004*, Pew Research Center, April 25, 2012, http://www.people-press.org/files/legacy-pdf/4-25-12%20Social%20Issues.pdf (accessed July 17, 2014)

FIGURE 8.4

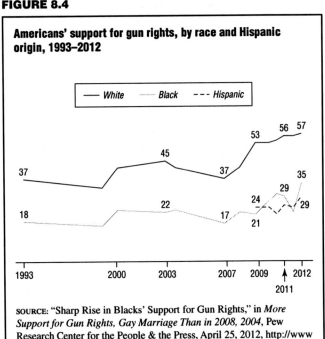

Americans' support for gun rights, by race and Hispanic origin, 1993–2012

SOURCE: "Sharp Rise in Blacks' Support for Gun Rights," in *More Support for Gun Rights, Gay Marriage Than in 2008, 2004*, Pew Research Center for the People & the Press, April 25, 2012, http://www.people-press.org/files/legacy-pdf/4-25-12%20Social%20Issues.pdf (accessed July 17, 2014)

broadly track differences in the prevalence of gun owner-ship. As Figure 8.3 shows, between 1993 and 2012 men were more likely to support gun rights than women, and as Figure 8.4 shows, during the same period whites have been more likely to support gun rights than African Americans and Hispanics.

Strictness of Gun Laws

Numerous polling organizations periodically conduct surveys assessing Americans' views on the strictness of gun control laws. Pew approaches the question of the strictness of gun control laws through the lens of gun control versus gun rights, conducting periodic surveys on

whether Americans believe it is more important to con-trol gun ownership or to protect the rights of gun owners. As Figure 8.5 shows, since 1993 the overall trend in opinion has been toward increased support for the rights of gun owners and decreasing support for gun control measures. In 2000, 66% of Americans favored gun con-trol and only 29% favored the rights of gun owners. By 2013, 50% of Americans favored gun control and 48% favored the rights of gun owners.

FIGURE 8.5

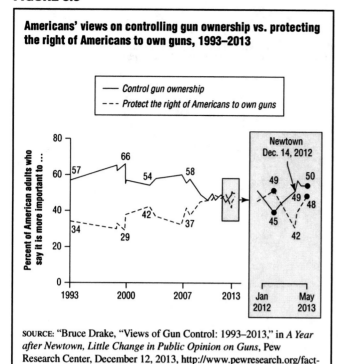

Americans' views on controlling gun ownership vs. protecting the right of Americans to own guns, 1993–2013

SOURCE: "Bruce Drake, "Views of Gun Control: 1993–2013," in *A Year after Newtown, Little Change in Public Opinion on Guns*, Pew Research Center, December 12, 2013, http://www.pewresearch.org/fact-tank/2013/12/12/a-year-after-newtown-little-change-in-public-opinion-on-guns/ (accessed July 17, 2014)

Support for the rights of gun owners accelerated sharply in 2008. (See Figure 8.5.) This increase in support is widely believed to have been triggered by the election of Barack Obama (1961–) as president, who was believed by many gun rights advocates to be hostile to their position. The NRA saw its membership grow during the Obama years and the organization's rhetoric in matters relating to President Obama was notably urgent. For example, in "Grassroots Activism Is a Powerful Force" (February 19, 2009, http://www.nraila.org/news-issues/articles/2009/president%60s-column-grassroots-activism.aspx), one of many postings on the website of the NRA's Institute for Legislative Action that cast the Obama administration as an unprecedented threat to Second Amendment rights, then president of the NRA John C. Sigler (1945–) reflected on the early days of the administration, "Everyone now knows the danger the Obama-Biden White House poses to the rights of law-abiding gun owners. We watched as President Barack Obama assembled his team from among the most anti-gun zealots to ever hold public office." As of November 2014, the organization still maintained this tone toward the Obama administration, both in its official publications and at public gatherings.

Frank Newport of the Gallup Organization reports in *Many Gun Owners Think Obama Will Try to Ban Gun Sales* (October 20, 2009, http://www.gallup.com/poll/123602/many-gun-owners-think-obama-will-try-ban-gun-sales.aspx) that ordinary gun owners shared the NRA's

sense of alarm. According to a Gallup survey from the first year of Obama's presidency, 55% of gun owners and 41% of Americans in general believed the new administration would attempt to ban the sale of guns in the United States. As correspondents from most major news outlets reported, this widespread belief led many gun owners to stockpile weapons and ammunition because of fears that they might soon be outlawed. For example, in early 2009 Ben Neary reported in "Fear of Regulation Drives Gun, Ammo Shortage" (USAToday.com, March 29, 2009) that "concern that the Obama administration could impose a new ban on some semiautomatic weapons is driving worried gun owners to stockpile ammunition and cartridge reloading components at such a rate that manufacturers can't meet demand."

Increased demand for guns and ammunition lasted throughout Obama's first term, and shortages became acute again in the wake of the December 2012 Sandy Hook shooting, which horrified the general public arguably more than any previous mass shooting, due to the fact that 20 of the 26 victims were six- and seven-year-old children. The effect of Sandy Hook on the gun control debate is illustrated in the enlarged portion of Figure 8.5, which shows that support for the rights of gun owners fell precipitously in December 2012. This change in public sentiment was fleeting, however, and as Obama and others called for increased gun control measures in the years that followed, the trends toward increased support for gun rights and increased demand for guns and ammunition continued. In "Record U.S. Gun Production as Obama 'Demonized' on Issue" (Bloomberg.com, February 20, 2014), Del Quentin Wilber quotes Dave Workman, the senior editor of *Gun Mag* (a magazine published by the gun rights group Second Amendment Foundation), who said, "Barack Obama is the stimulus package for the firearms industry. The greatest irony of the Obama administration is that the one industry that he may not have really liked to see healthy has become the healthiest industry in the United States."

The Gallup Organization provides further insight into the trend toward increased support for gun rights and increased opposition to gun control in the years following the Sandy Hook shooting. According to Rebecca Riffkin of Gallup, in *Americans' Dissatisfaction with Gun Laws Highest since 2001* (January 30, 2014, http://www.gallup.com/poll/167135/americans-dissatisfaction-gun-laws-highest-2001.aspx), in surveys conducted in January 2014 as part of Gallup's annual Mood of the Nation survey, the organization assessed public opinion on a range of issues including gun control. In response to the question of whether respondents were "very satisfied, somewhat satisfied, somewhat dissatisfied, or very dissatisfied" with "the nation's laws or policies on guns," a greater percentage of Americans (55%) claimed to be dissatisfied

FIGURE 8.6

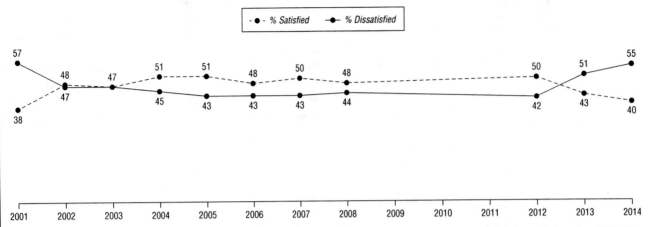

Americans' satisfaction with gun laws, 2001–14

NEXT, WE'D LIKE TO KNOW HOW YOU FEEL ABOUT THE STATE OF THE NATION IN EACH OF THE FOLLOWING AREAS. FOR EACH ONE, PLEASE SAY WHETHER YOU ARE—VERY SATISFIED, SOMEWHAT SATISFIED, SOMEWHAT DISSATISFIED, OR VERY DISSATISFIED. HOW ABOUT THE NATION'S LAWS OR POLICIES ON GUNS?

SOURCE: Rebecca Riffkin, "Americans' Satisfaction with Gun Laws and Policies," in *Americans' Dissatisfaction with Gun Laws Highest since 2001*, The Gallup Organization, January 30, 2014, http://www.gallup.com/poll/167135/americans-dissatisfaction-gun-laws-highest-2001.aspx (accessed July 17, 2014). Copyright © 2014 Gallup, Inc. All rights reserved. The content is used with permission; however, Gallup retains all rights of republication.

FIGURE 8.7

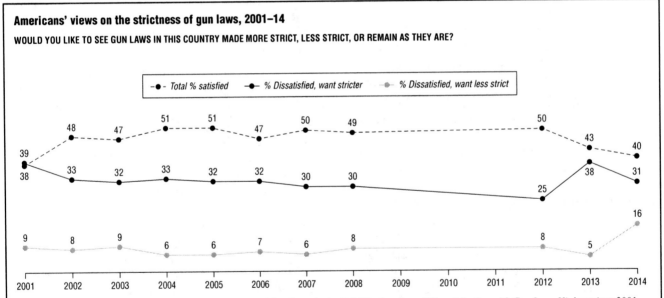

Americans' views on the strictness of gun laws, 2001–14

WOULD YOU LIKE TO SEE GUN LAWS IN THIS COUNTRY MADE MORE STRICT, LESS STRICT, OR REMAIN AS THEY ARE?

SOURCE: Rebecca Riffkin, "Satisfaction with the Strictness of Gun Laws in the U.S.," in *Americans' Dissatisfaction with Gun Laws Highest since 2001*, The Gallup Organization, January 30, 2014, http://www.gallup.com/poll/167135/americans-dissatisfaction-gun-laws-highest-2001.aspx (accessed July 17, 2014). Copyright © 2014 Gallup, Inc. All rights reserved. The content is used with permission; however, Gallup retains all rights of republication.

with the nation's gun laws than at any point since 2002. (See Figure 8.6.) This figure was up dramatically since 2012, when 42% of U.S. residents expressed dissatisfaction and 50% expressed satisfaction with the gun law status quo—levels of dissatisfaction and satisfaction that had been roughly steady since 2004.

Besides a spike in dissatisfaction with the nation's gun laws, Riffkin notes that there had been a shift in the nature of that dissatisfaction since 2012. As Figure 8.7 shows, between 2013 and 2014 there was a dramatic rise in the percentage reporting dissatisfaction because they felt the current laws were too strict. The proportion of the population that wanted less strict gun laws had been relatively flat since 2001, fluctuating between 5% and 9%, but in 2014, 16% of respondents claimed to want less strict laws. By contrast, the high proportion of respondents answering in 2001 that they were dissatisfied with the nation's gun laws was largely driven by the percentage of respondents (39%) who believed gun laws were not strict enough. The period between 2012 and 2014 also saw a substantial increase in the percentage of people reporting they were "very dissatisfied" as opposed to "somewhat dissatisfied" with the nation's gun laws and policies. (See Figure 8.8.) In 2012, 23% of respondents claimed to be "very dissatisfied" and 19% "somewhat dissatisfied." These levels were consistent with survey responses over the course of the preceding 10 years. However, in 2014 a record 35% of Americans claimed to be "very dissatisfied"

with the gun law status quo. Riffkin links these changes to increased perceptions of the threat to gun rights in the aftermath of Sandy Hook, as well as to Obama's inclusion, in his January 2014 State of the Union speech, of his desire to see new gun control measures passed.

Harris Interactive presents a slightly different picture of public opinion on the question of whether gun laws are too strict or not strict enough. As shown in Table 8.1, the overall support for stricter gun control laws fell between 1998 and 2014, but this decrease (from 69% in 1998 to 51% in 2014) was not accompanied by a corresponding rise in the percentage of respondents claiming to want

FIGURE 8.8

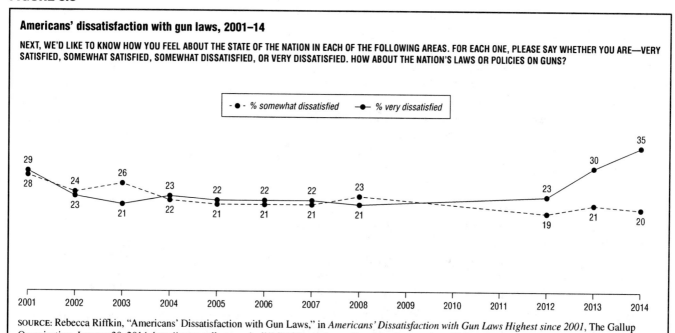

Americans' dissatisfaction with gun laws, 2001–14

NEXT, WE'D LIKE TO KNOW HOW YOU FEEL ABOUT THE STATE OF THE NATION IN EACH OF THE FOLLOWING AREAS. FOR EACH ONE, PLEASE SAY WHETHER YOU ARE—VERY SATISFIED, SOMEWHAT SATISFIED, SOMEWHAT DISSATISFIED, OR VERY DISSATISFIED. HOW ABOUT THE NATION'S LAWS OR POLICIES ON GUNS?

TABLE 8.1

Americans' views on stricter vs. less strict gun control laws, by year, political party, and gender, selected years 1998–2014

[Base: All adults]

"IN GENERAL, WOULD YOU SAY YOU FAVOR STRICTER GUN CONTROL OR LESS STRICT GUN CONTROL?"

| | 1998 | 1999 | 2000 | 2004 | 2008 | 2010 | 2014 | Political party | | | Gender | |
| | | | | | | | | Rep. | Dem. | Ind. | Male | Female |
	%	%	%	%	%	%	%	%	%	%	%	%
Stricter	69	63	63	52	49	45	51	30	76	47	47	55
Less strict	23	25	28	22	20	26	17	30	3	20	20	14
Neither*	7	10	6	20	21	20	23	30	14	25	26	21
Don't know/refused	1	2	4	7	10	10	8	10	7	7	7	10

*There was a change in the question in 2004. In the previous surveys "neither" was not offered as a possible response but was accepted if given. In this new survey it was offered as a possible response.
Note: Percentages may not add to 100% because of rounding

less strict laws. An especially notable difference in responses became detectable in 2004, when the group changed its question format. Prior to that time, it offered respondents the choice of answering that gun laws should be "stricter" or "less strict," but it accepted an answer of "neither" when respondents gave that answer. Starting in 2004, Harris explicitly offered "neither" as a possible answer. As a result, markedly fewer respondents answered that gun laws should be stricter and roughly equal proportions answered that they should be less strict or neither. Harris found a noticeable additional decline in support for stricter gun laws between 2004 (when 52% supported this option) and 2010 (when 45% supported this option), but by 2014 support for the stricter option had returned to 51%.

Harris finds predictable differences of opinion about the strictness of gun control laws by political affiliation and gender in 2014. (See Table 8.1.) Three-quarters (76%) of Democrats favored stricter gun laws, whereas Republicans were split equally among those who wanted stricter laws (30%), less strict laws (30%), and neither (30%). Women (55%) were more likely than men (47%) to favor stricter laws, and men were more likely than women to favor less strict laws (20% versus 14%, respectively) as well as neither (26% versus 21%, respectively).

Support for a ban on handguns, one of the strictest forms of gun control laws that has ever been proposed in the United States, has generally fallen over time since the middle of the 20th century. Such bans have been imposed at the municipal level but never at the state or federal level, and in 2008 the U.S. Supreme Court ruling in *District of Columbia v. Heller* (554 U.S. 570 [2008]), which struck down the District of Columbia's ban on the possession of handguns, made it increasingly unlikely that such policies will be imposed in the future. As Table 8.2 shows, support for a handgun ban fell steadily over the course of the late 20th and early 21st centuries, from 60% in 1959 to 25% in 2013.

Gun Control and Mass Shootings

Support for increased gun control tends to spike in the aftermath of mass shootings, but such support has typically been fleeting. Part of the reason that increased restrictions on gun ownership have not been passed at the federal level following tragedies such as Sandy Hook is that Americans are divided on the question of whether access to guns is a primary factor in such shootings. Table 8.3 shows the results of two separate Gallup surveys, one conducted in January 2011 and one conducted in September 2013, both of which assess respondents' views on the factors that are most to blame for school shootings. In both 2011 and 2013, 48% believed the "failure of the mental health system to identify individuals who are a danger to others" deserved a "great deal"

TABLE 8.2

Americans' views on a potential handgun ban, selected years 1959–2013

DO YOU THINK THERE SHOULD OR SHOULD NOT BE A LAW THAT WOULD BAN THE POSSESSION OF HANDGUNS, EXCEPT BY THE POLICE AND OTHER AUTHORIZED PERSONS?

	Yes, should be	No, should not be	No opinion
	%	%	%
2013 Oct 3–6	25	74	2
2012 Dec 19–22	24	74	2
2011 Oct 6–9	26	73	1
2010 Oct 7–10	29	69	1
2009 Oct 1–4	28	71	1
2008 Oct 3–5	29	69	2
2007 Oct 4–7	30	68	2
2006 Oct 9–12	32	66	2
2005 Oct 13–16	35	64	1
2004 Oct 11–14	36	63	1
2003 Oct 6–8	32	67	1
2002 Oct 14–17	32	65	3
2000 Aug 29–Sep 5	36	62	2
1999 Apr 26–27	38	59	3
1999 Feb 8–9	34	64	2
1993 Dec 17–19	39	60	1
1993 Mar 12–14	42	54	4
1991 Mar 21–24	43	53	4
1990 Sep 10–11	41	55	4
1988 Jul 1–7	37	59	4
1987 Oct 12–18	42	50	8
1982 Apr 2–5	41	54	5
1981, Jun 19–22	41	54	5
1981 Apr 3–6	39	58	3
1980 Dec 12–21	38	51	11
1979 Nov 2–5[a]	31	65	4
1975 Mar 7–10[a]	41	55	4
1965 Jan 7–12[b]	49	44	7
1959 Jul 23–28[b]	60	36	4

[a]Wording: Now here is a question about pistols and revolvers. Do you think there should or should not be a law which would forbid the possession of this type of gun except by the police and other authorized persons?
[b]Wording: What about the possession of pistols and revolvers—do you think there should be a law which would forbid the possession of this type of gun except by the police or other authorized persons?

SOURCE: "Do You Think There Should or Should Not Be a Law That Would Ban the Possession of Handguns, Except by the Police and Other Authorized Persons?" in *Guns*, The Gallup Organization, 2014, http://www.gallup.com/poll/1645/guns.aspx (accessed July 14, 2014). Copyright © 2014 Gallup, Inc. All rights reserved. The content is used with permission; however, Gallup retains all rights of republication.

of blame for mass shootings, and a large number of the remaining respondents (30% in 2011 and 32% in 2013) believed this factor deserved a "fair amount" of blame. A lower proportion of respondents in both surveys (46% in 2011 and 40% in 2013) felt that "easy access to guns" deserved a "great deal" of blame for mass shootings, and the remaining respondents were relatively evenly split between those who believed access to guns deserved a "fair amount" of blame, "not much" blame, or no blame at all. According to Gallup, then, in 2013 only 61% of Americans believed access to guns had either a "great deal" or a "fair amount" to do with mass shootings, whereas 80% believed problems with the mental health system had either a "great deal" or a "fair amount" to do with such incidents.

TABLE 8.3

Americans' views on factors to blame for school shootings, 2011 and 2013

THINKING ABOUT MASS SHOOTINGS THAT HAVE OCCURRED IN THE UNITED STATES IN RECENT YEARS, FROM WHAT YOU KNOW OR HAVE READ, HOW MUCH DO YOU THINK EACH OF THE FOLLOWING FACTORS IS TO BLAME FOR THE SHOOTINGS—A GREAT DEAL, A FAIR AMOUNT, NOT MUCH, OR NOT AT ALL? HOW ABOUT—[RANDOM ORDER]?

	Great deal %	Fair amount %	Not much %	Not at all %	No opinion %
Easy access to guns					
2013 Sep 17–18	40	21	16	20	3
2011 Jan 14–16*	46	19	16	18	1
Violence in movies, video games, and music lyrics					
2013 Sep 17–18	32	24	23	20	2
2011 Jan 14–16*	31	27	26	15	2
The spread of extremist viewpoints on the Internet					
2013 Sep 17–18	29	28	22	15	6
2011 Jan 14–16*	32	31	19	12	6
Failure of the mental health system to identify individuals who are a danger to others					
2013 Sep 17–18	48	32	11	8	2
2011 Jan 14–16*	48	30	11	8	3
Inflammatory language from prominent political commentators					
2013 Sep 17–18	18	19	30	28	6
2011 Jan 14–16*	19	25	29	25	3
Drug use					
2013 Sep 17–18	37	29	17	15	3
2011 Jan 14–16*	42	35	16	6	2
Insufficient security at public buildings including businesses and schools					
2013 Sep 17–18	29	29	26	14	2

*Asked of a half sample.

SOURCE: "Thinking about Mass Shootings That Have Occurred in the United States in Recent Years, from What You Know or Have Read, How Much Do You Think Each of the Following Factors Is to Blame for the Shootings—a Great Deal, a Fair Amount, Not Much, or Not at All?" in *Guns*, The Gallup Organization, 2014, http://www.gallup.com/poll/1645/guns.aspx (accessed July 14, 2014). Copyright © 2014 Gallup, Inc. All rights reserved. The content is used with permission; however, Gallup retains all rights of republication.

TABLE 8.4

Americans' views on best approach to preventing future school shootings, 2013

IF YOU HAD TO CHOOSE, WHICH APPROACH TO PREVENTING FUTURE SCHOOL SHOOTINGS SHOULD THE PRESIDENT AND CONGRESS FOCUS ON—[ROTATED: MAKING MAJOR CHANGES TO THE LAWS ON THE SALE OF GUNS AND AMMUNITION, (OR) MAKING MAJOR CHANGES TO SCHOOL SECURITY MEASURES AND THE MENTAL HEALTH SYSTEM]?

	Change laws on guns and ammunition	Change school security measures and mental health system	No opinion
2013 Jan 19–20	30%	65%	6%

SOURCE: "If You Had to Choose, Which Approach to Preventing Future School Shootings Should the President and Congress Focus on— [ROTATED: Making Major Changes to the Laws on the Sale of Guns and Ammunition, (or) Making Major Changes to School Security Measures and the Mental Health System]?" in *Guns*, The Gallup Organization, 2014, http://www.gallup.com/poll/1645/guns.aspx (accessed July 14, 2014). Copyright © 2014 Gallup, Inc. All rights reserved. The content is used with permission; however, Gallup retains all rights of republication.

In another Gallup survey, which was conducted in January 2013, respondents were asked to choose between two choices for preventing future school shootings: "making major changes to the laws on the sale of guns and ammunition" or "making major changes to school security measures and the mental health system." Only 30% of respondents believed changes to gun and ammunition laws would be the best approach, whereas 65% believed changes to school security and the mental health system were the best approach. (See Table 8.4.)

CHAPTER 9

THERE SHOULD BE STRICTER GUN CONTROL LAWS

This chapter presents a sample of the arguments used by the proponents of strong federal gun control to support their position. Chapter 10 provides arguments put forward by opponents of strong federal gun control.

TESTIMONY OF MICHAEL A. NUTTER, MAYOR OF PHILADELPHIA AND PRESIDENT OF THE UNITED STATES CONFERENCE OF MAYORS, BEFORE THE U.S. SENATE COMMITTEE ON THE JUDICIARY ON THE TOPIC OF THE PROPOSED ASSAULT WEAPONS BAN OF 2013, FEBRUARY 27, 2013

Senator Feinstein, just as gun violence has been personal for you, it has been personal for me. The first police officer my city lost after I became Mayor was killed with an assault weapon when he responded to a bank robbery in a supermarket on a Saturday morning. A 12-year veteran of the Philadelphia Police Department, Sergeant Stephen Liczbinski was 39 years old. He left a wife and three children. Neither our police officers nor our citizens, and especially our children, should be confronted with these weapons on the streets of our cities, in our schools, in our movie theaters, in our shopping malls, in our places of worship, or in other civilian settings.

Again and again and again, Americans have been stunned by senseless acts of violence involving assault weapons and large-capacity magazines: Columbine (April 1999, 13 murdered), Virginia Tech (April 2007, 32 murdered), Tucson (January 2011, 6 murdered, 12 wounded including one Congresswoman), Aurora (July 2012, 12 murdered), Oak Creek (August 2012, 6 murdered). The December 14 tragedy at Sandy Hook Elementary which killed 20 young children and six educators in Newtown remains incomprehensible to us all. Too many times during the last year, mayors have expressed shock at a mass shooting.

Even more frequently, many of us must cope with the gun violence that occurs on the streets of our cities. Citizens have been killed on Philadelphia's streets by handguns with high capacity magazines as well as by assault rifles. To me and to America's mayors these are weapons of mass destruction and they are destroying our communities, our streets and our families.

The U.S. Conference of Mayors has been calling for sensible gun laws to protect the public for more than 40 years. Our call for a ban on assault weapons dates back to 1991. Mayors and police chiefs from cities of all sizes have worked together in this effort over the years.

Gun Violence in Cities

We have done that because of the tremendous toll gun violence takes on the American people day in and day out:

- Every year in America more than 100,000 people are shot, and 31,537 of them die, including 11,583 who are murdered.

- Every year, 18,000 children and teens are shot, and 2,829 of them die, including 1,888 who are murdered.

- Every day in America, 282 people are shot and 86 of them die, including 32 who are murdered.

- Every day 50 children and teens are shot and eight of them die, including five who are murdered.

Gun violence disproportionately affects urban areas. Our nation's 50 largest metro areas have 62 center cities, and these cities account for 15 percent of the population, but 39 percent of gun-related murders and 23 percent of total homicides.

Philadelphia, like many major cities, has struggled to control gun violence for years. However, despite our recent success at employing more effective policing techniques, deaths due to gun violence have not fallen. Let me use one set of statistics to illustrate this point:

Last year, the number of shooting victims in Philadelphia was 1,282. This is down considerably from the year before—and was the lowest number since we began tracking shooting victims in the year 2000. However, the number of homicides was up slightly—331, seven more than the previous year. How are these two statistics possible? The answer is that the homicide victims have more bullets killing them. Or, to put it another way, there are more rounds being fired and more intentional head shots. So despite better policing, when someone in Philadelphia is shot, they are now more likely to die.

I would note that Pennsylvania does not have stringent gun restrictions. When the City of Philadelphia adopted strict gun laws a few years ago, the state supreme court struck those laws down. This is why we need federal legislation. Cities alone cannot reduce gun violence. We are doing everything that we can, but we are still losing the battle thanks to the proliferation of guns in this nation.

Philadelphia's story is not unique. Mayors everywhere struggle with gun violence, using scarce city resources to fight it—resources which we should be using to educate our children, create jobs for our residents, and revitalize our cities.

I have with me this morning a letter originally sent just three days after the Newtown tragedy occurred and now signed by 212 mayors which calls on the President and Congress to take immediate action and make reasonable changes to our gun laws and regulations. Listed first among our recommended changes is enactment of legislation to ban assault weapons and high-capacity magazines, your legislation Senator Feinstein. I ask that you include this letter in the record of this hearing.

The Assault Weapons Ban Does Not Violate the Second Amendment

Since the shootings in Newtown, the question, "If not now, when?" has been raised often in news media columns, editorials, and other arguments for swift and meaningful action to combat gun violence. And well it should, because it's the right question. For too many in the gun rights community, however, the answer to the question is always "never," and the reason is always the Second Amendment's protection of the "right to bear arms."

Harvard University's Lawrence Tribe, one of the nation's most respected experts on Constitutional law, told this Committee just a few weeks ago that, after examining the various proposals being considered—including the ban on assault weapons and high-capacity magazines—"I am convinced that nothing under discussion in the Senate Judiciary Committee represents a threat to the Constitution or even comes close to violating the Second Amendment or the Constitution's structural limits either on congressional power or on executive authority."

The 1994 Assault Weapons Ban Worked

In addition we've been told by the gun rights community that the assault weapons ban didn't work before and it won't work now. Research shows that the 1994 ban did work and that since it expired the use of assault weapons by criminals has increased:

- A Justice Department study of the 1994 assault weapons ban found that it was responsible for a 6.7 percent decrease in total gun murders, holding all other factors equal. That study also found that "assault weapons are disproportionately involved in murders with multiple victims, multiple wounds per victim, and police officers as victims."

- An updated assessment of the federal ban found that the use of assault weapons in crime declined by more than two-thirds about nine years after 1994 Assault Weapons Ban took effect.

- A recent study by the Violence Policy Center found that, between 2005 and 2007, one in four law enforcement officers slain in the line of duty was killed with an assault weapon.

- The Police Executive Research Forum reports that 37 percent of police departments reported seeing a noticeable increase in criminals' use of assault weapons since the ban expired.

It's Time to Pass the Assault Weapons Ban

Mayors consider protecting the safety of their citizens and their cities their highest responsibility. We know that keeping our cities and our citizens safe requires more than passing sensible gun laws, including the assault weapons ban, but we also know that we cannot keep our cities safe unless we pass such laws. The Assault Weapons Ban of 2013 would:

- Prohibit the sale, manufacture, transfer and importation of 157 of the most commonly-owned military-style assault weapons and ban an additional group of assault weapons that can accept a detachable ammunition magazine and have one or more military characteristics;

- Ban large-capacity magazines and other ammunition feeding devices that hold more than 10 rounds of ammunition—devices which allow shooters to fire numerous rounds in rapid succession without having to stop and reload; and

- Protect the rights of law-abiding citizens who use guns for hunting, household defense or legitimate recreational purposes and exempt all assault weapons lawfully possessed at the date of enactment from the ban.

- Require background checks on all future transfers of assault weapons covered by the legislation, including sale, trade and gift;

- Require that grandfathered assault weapons be stored safely using a secure gun storage or safety device in order to keep them away from prohibited persons; and

- Prohibit the sale or transfer of high-capacity ammunition feeding devices currently in existence.

This is common sense legislation which will help us to reduce the number of people, including police officers, who are shot and killed in our cities and throughout our nation. This legislation deserves a vote. This legislation deserves to be passed, by this Committee, by the Senate, and by the House so that the President can sign it into law.

I know it will take an act of political courage for many Members of Congress to support the Assault Weapons Ban of 2013, but the time for such political courage is now. How many more children, how many more police officers do we have to lose for our elected representatives to do the right thing? The nation's mayors pledge to work with you to build a safer America for our children and all of our citizens. (http://www.judiciary.senate.gov/imo/media/doc/2-27-13NutterTestimony.pdf)

EXCERPT OF WRITTEN TESTIMONY SUBMITTED BY THE AMERICAN ACADEMY OF PEDIATRICS TO THE U.S. SENATE COMMITTEE ON THE JUDICIARY, SUBCOMMITTEE ON THE CONSTITUTION, CIVIL RIGHTS, AND HUMAN RIGHTS HEARING "PROPOSALS TO REDUCE GUN VIOLENCE: PROTECTING OUR COMMUNITIES WHILE RESPECTING THE SECOND AMENDMENT," FEBRUARY 12, 2013

The American Academy of Pediatrics (AAP) [is] a non-profit professional organization of more than 60,000 primary care pediatricians, pediatric medical sub-specialists, and pediatric surgical specialists dedicated to the health, safety, and well-being of infants, children, adolescents, and young adults.

The AAP is committed to protecting children from the horrific consequences of gun violence and traumatic events, and ensuring children's safety within their homes, schools and communities. The tragedy at Sandy Hook Elementary School in Newtown, CT, serves as a stark reminder that gun violence affects communities nationwide. Unfortunately, while outbursts of mass violence like that at Sandy Hook are at least relatively rare, the scourge of gun violence is a phenomenon that our nation's children experience every single day. In 2008 and 2009, 5,740 children were killed by guns, meaning that 55 died each week during that period. The causes of gun violence are varied and complex but we must act to develop a comprehensive response centered on the rights [of] children and families to be safe and free from its harmful effect in their lives and within their communities.

A Public Health Approach to Reducing Gun Violence

Gun violence is a public health issue with particularly pernicious effects on children. Firearm related deaths continue to be one of the top three causes of death among American youth, causing twice as many deaths as cancer, five times as many as heart disease, and 15 times as many as infections. In 2009, 84.5 percent of all homicides of people 15 to 19 years of age were firearm-related. The United States has the highest rates of firearm-related death (including homicide, suicide and unintentional deaths) among high income countries. For youth ages 15 to 24 years of age, [firearm] homicide rates were 35.7 times higher than in other high income countries. For over 20 years, the AAP has supported stronger gun violence prevention policies because of the public health implications of this problem. Reducing its impact must be consistent with other initiatives that have reduced injury and mortality through evidence-based prevention efforts.

Policy of the AAP, based on extensive research, is that absence of guns from children's homes and communities is the most reliable and effective measure to prevent firearm-related injuries in children and adolescents. Access to a firearm increases the risk of unintended injury or death among all children. A gun stored in the home is associated with a threefold increase in the risk of homicide and a fivefold increase in the risk of suicide. Individuals possessing a firearm are more than four times more likely to be shot during an assault than those who do not own one. The association of a gun in the home and increased risk of suicide among adolescents is well-documented, even among teens with no underlying psychiatric diagnosis. These health risks associated with gun violence point toward the need for long term research investments on effective strategies to protect children and adolescents, particularly those within at-risk communities.

As part of its engagement with the White House Taskforce on Gun Violence Prevention, the AAP recommended federal support for gun violence research, and is pleased the President's plan recommended $10 million to support Centers for Disease Control Prevention (CDC) research into the causes and prevention of gun violence; $10 million for CDC to conduct further research regarding the relationship between video games, media images and violence; and $20 million to expand the National Violent Death Reporting System from 18 to 50 states. The AAP urges Congress to support these efforts within the annual appropriations process and to eliminate any restrictive language that may discourage gun violence research.

The AAP supports policies aimed at protecting children and adolescents from the destructive effect[s] of guns through strong gun safety legislation that bans assault weapons and high-capacity magazines, requires universal background checks, and mandates safe firearm storage. Consistent with this policy, the AAP has endorsed the *Assault Weapons Regulatory Act of 2013* (S. 150/H.R. 437). According to a recent analysis by the Violence Policy Center (VPC), the five states (Alabama, Alaska, Louisiana, Montana, and Wyoming) with the least restrictive gun laws and high gun ownership rates also had the highest per capita gun death rates. States with strong gun laws and low rates of gun ownership had far lower rates of firearm-related death. The AAP is encouraged that bipartisan efforts are underway to strengthen gun laws, and looks forward to the opportunity to review those plans as they materialize.

Responsible Gun Ownership

Policies to support safe and responsible ownership can go a long way toward keeping firearms out of the hands of children and adolescents who may harm themselves or others. An estimated 57 million Americans own[ed] 283 million firearms in 2004. Among gun owners with a child 18 years old or younger, 31 percent store their guns unlocked, 21 percent store them loaded, and 8.3 percent store them unlocked and loaded. Safe gun storage can reduce the risk of youth injury and suicide by more than 70 percent; therefore, efforts to educate families and require responsible practices through child access prevention (CAP) laws should be supported as important, but common sense, interventions.

CAP laws impose criminal liability on adults who negligently leave firearms accessible to children or otherwise allow children access to firearms. One study found that in twelve states where such laws had been in effect for at least one year, unintentional firearm deaths fell by 23% from 1990–94 among children under 15 years of age. Laws reducing child access also are associated with lower overall adolescent suicide.

The AAP commends the Obama Administration's safe gun storage campaign proposal and urges Congress to support this initiative. Medical professionals and law enforcement officials should play an important role in implementing this campaign. The AAP's *Bright Futures* clinical guidance recommends that pediatricians ask about guns in the home and that they provide age-appropriate safety counseling, similar to the guidance they provide on other injury risks, like drowning and parental tobacco use. Physician counseling of parents about firearm safety, particularly when combined with the distribution of gun locks, has been demonstrated to be an effective prevention measure and shown to increase compliance with safe storage principles.

At the federal level, the Affordable Care Act includes language barring the Secretary of the Department of Health and Human Services and health plans participating in the exchanges from collecting and housing information regarding the presence of firearms in the home. The AAP welcomes the president's guidance that the Affordable Care Act does not prohibit physicians from counseling patients regarding firearms.

The AAP remains concerned about state efforts to infringe upon physicians' rights to provide this crucial counsel such as the *Firearm Owners' Privacy Act*, enacted in Florida, which prevented physicians from providing such counsel under threat of financial penalty and potential loss of licensure. The law has been blocked from implementation by a U.S. District court but similar policies have been introduced in seven other states: Alabama, Minnesota, North Carolina, Oklahoma, South Carolina, Tennessee and West Virginia. This right must be protected to mitigate risk of injury to children in the environments in which they live and play. (http://www.judiciary.senate.gov/imo/media/doc/021213RecordSubmission-Durbin.pdf)

WRITTEN TESTIMONY OF DANIEL W. WEBSTER, SCD, MPH, PROFESSOR AND DIRECTOR OF THE JOHNS HOPKINS CENTER FOR GUN POLICY AND RESEARCH, BEFORE THE U.S. SENATE COMMITTEE ON THE JUDICIARY FOR THE HEARING "WHAT SHOULD AMERICA DO ABOUT GUN VIOLENCE?" JANUARY 30, 2013

On January 14–15, 2013, more than twenty of the top researchers and gun policy experts gathered to participate in a Summit on Reducing Gun Violence in America at Johns Hopkins, and presented findings and analyses that were just published in a book. These leading scholars identified numerous weaknesses in current federal firearms policy which enable criminals, those with severe mental illness, perpetrators of domestic violence, and underage youth to obtain firearms. These weaknesses in our firearms policies play an important role in explaining why the United States' homicide rate is seven times higher than the average rate among other high-income countries.

A recent national survey we conducted found very broad support—among gun owners and non-gunowners and across [various] political party affiliation—for laws prohibiting these and other high-risk groups from possessing firearms. There was similarly broad support for measures to keep guns from these groups, such as requiring background checks for all gun sales and stronger laws governing licensed gun dealers. Importantly, research shows that prohibiting high-risk groups from possessing firearms reduces violence and saves lives, especially if necessary records are available for law enforcement to deny prohibited individuals.

Opponents of stronger gun laws often claim that we simply need to do a better job of enforcing current gun laws. But current federal laws are written in ways that make it very difficult to hold firearm sellers, whether licensed dealers or private sellers, accountable if they sell firearms to criminals or traffickers. Non-licensed sellers of firearms have no obligation to ensure that the prospective purchasers have passed a background check and can legally possess firearms.

Such a policy is indefensible and is commonly exploited by criminals and traffickers. It is not surprising that nearly eighty percent of handguns used by offenders incarcerated in state prisons report that they acquired their handguns from non-licensed sellers—friends, family, and sellers in the underground market. Nor is it surprising that states that fail to regulate private handgun transactions export guns to criminals in states that do regulate private handgun sales. If you follow the logic of arguments that requiring background checks for private gun sales is pointless because criminals won't obey the law, then laws against drunk driving are pointless because drunks will always disobey those laws. Just as drunk driving laws provide law enforcement with the tools to arrest individuals who break those laws and deter others from driving drunk, requiring background checks for all sales will provide law enforcement with the tools it needs to combat illegal gun trafficking and keep guns from prohibited individuals.

Unfortunately, Congress has enacted several laws that shield scofflaw gun dealers from scrutiny, civil penalties, and criminal prosecution. The 1986 Firearm Owners Protection Act weakened penalties for gun sales violations, increased standards of proof for prosecutions and actions against licensed gun dealers, and limited ATF law compliance inspections. The Protection of Lawful Commerce in Arms Act provided special immunity from lawsuits for negligent practices which enable criminals and other prohibited individuals to obtain guns. The Tiahrt amendments provided further protections to licensed gun dealers who sell many guns that subsequently are recovered from criminals.

There is a growing body of research that has consistently demonstrated that laws which increase gun seller accountability and increase the risk to those involved in illegal gun transactions significantly reduce the number of guns diverted for criminal use. Whereas the federal Tiahrt amendments have been shown to increase the diversion of guns to criminals from suspect gun dealers, strong regulation and oversight of gun dealers reduces guns diverted to criminals, as does being vulnerable to lawsuits for making illegal sales. Research has also shown that regulation of private sales of handguns, mandatory reporting of loss or theft of firearms from private owners, and permit-to-purchase licensing for handguns reduces the diversion of guns to criminals.

By adopting many laws shown to be effective at that the state level, Congress could significantly reduce the availability of guns to dangerous individuals, which would translate into fewer lives lost, safer streets and homes, increased quality of life, and reduced government expenditures on health care, disability payments, criminal justice, and corrections. (http://www.judiciary.senate.gov/download/01-30-13-record-submission_-feinstein&download=1)

WRITTEN TESTIMONY OF CHIEF JIM BUEERMANN (RET.), PRESIDENT, POLICE FOUNDATION, WASHINGTON, D.C., BEFORE THE U.S. SENATE COMMITTEE ON THE JUDICIARY FOR THE HEARING "WHAT SHOULD AMERICA DO ABOUT GUN VIOLENCE?" JANUARY 30, 2013

I write to you in my capacity as both President of the Police Foundation and the former Chief of Police of the Redlands, CA Police Department. The Police Foundation, established in 1970 by the Ford Foundation, is a non-partisan, non-constituency research organization. Our mission is to advance policing through innovation and scientific research. The Foundation is committed to disseminating science and evidence-based practices to the field. My written testimony reflects these principles and my personal experience after 33 years as a police officer during which time I witnessed countless acts of violence. I urge the passage of the Assault Weapons Ban Act of 2013 and ask Congress to consider funding additional scientific research to help this country implement evidence-based approaches to reducing gun violence in our communities and schools.

The most recent available data reveal this alarming picture of America's experience with gun-related violence: in 2011, of the 32,163 deaths from firearms, 19,766 were suicides and 11,101 were homicides. Additionally, there were 467,321 non-fatal violent crimes committed with a firearm. These numbers all reflect the unique position of the United States in relation to other high-income nations: our homicide rate is 6.9 times higher than the combined homicide rate of 22 other high-income countries. We all know that gun violence must be stemmed. The Police Foundation supports a comprehensive and holistic approach to preventing and reducing gun violence that includes:

- Legislation that bans assault weapons, requires universal background checks for all firearm purchases and limits high capacity ammunition feeding devices to ten rounds;

- Enhanced funding for research on the availability of firearms, the causes and prevention of gun violence and the connection between mental health and gun violence;

- Specific funding to replicate the 1996 US DOJ, National Institute of Justice study *Guns in America* that provided a comprehensive view of guns in our society;

- Increased funding to states for community-based mental health treatment; and,

- Sustained funding and support of the Justice and Mental Health Collaboration Program Act, which allows for collaborative efforts between law enforcement, criminal justice and mental health professionals.

Gun violence, especially violence that is mental health-related, is a complex social, cultural, health and safety issue. It is one that we do not know enough about. As the leader of a research organization that focuses on policing crime and disorder, I stress the need for scientific research and an evidence-based approach to understanding important societal issues. As a country, we need a robust and rigorous agenda on the causes of gun violence, effective, community-based prevention and intervention strategies and the link between mental illness and gun violence. Lifting the freeze on gun violence research at the Centers for Disease Control [and Prevention] is heartening, and I hope Congress will support additional funding for interdisciplinary, scientific research and collaboration across government agencies, including the Department of Justice and the Department of Health and Human Services.

Mental health-related gun violence has been brought to the fore with the shootings in Newtown, CT, Aurora, CO, and Tucson, AZ. While these tragic incidents are statistically rare, when combined with the number of gun-related suicides each year, the necessity of addressing the mental health needs of individuals, and the availability of firearms in our communities, is paramount.

We do not want to stigmatize individuals with mental illness nor solely focus the current dialogue on gun violence on the role of mental illness. The best available data on violence attributable to mental illness shows that 3-5% of violent acts are committed by individuals with mental illness and most of these acts do not involve guns. Yet, we cannot ignore the number of gun-involved suicides each year and the connection between mass shootings and mental illness. Increased scientific research across the fields of medicine, public health, criminal justice and law will help us understand how to prevent mental health-related gun violence. This requires both robust funding and time.

As a former chief of police, I recognize that local law enforcement agencies require immediate strategies to prevent another incident of mass violence. Earlier this month, the Police Foundation convened a roundtable meeting of expert researchers and practitioners from the fields of law enforcement, mental health, public health, criminal justice and policy. The group discussed how available interdisciplinary research might be used to develop practical strategies for law enforcement that prevent mental health-related gun violence. Existing research establishes the difficulty in predicting a violent act, but the group committed to three strategies that law enforcement can adopt now. Based on innovative practices defined in the literature, the group proposed that law enforcement executives:

- Create local partnerships with mental health service providers, school officials and appropriate community groups to develop a mental health crisis response capacity;

- Advocate for increased mental health services in their communities. Law enforcement executives should convene local service providers and community members to assess local mental health services and community needs and increase community members' knowledge of the exiting science on mental health and gun violence;

- Adopt specific policies and practices that reduce the availability of guns to people in mental health crisis, institutionalize mental health training for their officers and facilitate community-wide "mental health first aid" training for all community members.

Clearly, more work needs to be done in this area so police departments can effectively operationalize these ideas. With additional Congressional support, strategies like these can be supported by legislation such as the Justice and Mental Health Collaboration Act or through an enhancement of programs at the Department of Justice and the Departments of Health and Human Services and Education. The JMHC Act has bipartisan support across the House of Representatives and Senate, and I ask that Congress sustain funding for these important ideas as part of a targeted approach to specifically reducing gun violence.

Charting a path to respond to gun violence will not be easy, but I encourage Congress to rely on the police, community leaders and science to guide that path. The Police Foundation, along with law enforcement leaders across the country, support reducing the availability of assault weapons and high capacity ammunition feeding device[s] as a first step to reducing gun violence. However, to effectively reduce gun violence, there must be more comprehensive action. Congress should prioritize funding to better understand guns in America, research on the causes and prevention of gun violence and the connection between mental illness and gun violence. It should also enhance the funding and availability of mental health services in communities, and support programs that increase local collaboration between law enforcement, criminal justice and mental health professionals. (http://www.judiciary.senate.gov/download/01-30-13-record-submission_-feinstein&download=1)

CHAPTER 10
THERE SHOULD NOT BE STRICTER GUN CONTROL LAWS

This chapter presents a sample of the arguments used by the opponents of strong federal gun control to support their position. Chapter 9 provides arguments put forward by supporters of strong federal gun control.

TESTIMONY OF WAYNE LAPIERRE, EXECUTIVE VICE PRESIDENT OF THE NATIONAL RIFLE ASSOCIATION OF AMERICA, BEFORE THE U.S. SENATE COMMITTEE ON THE JUDICIARY FOR THE HEARING "WHAT SHOULD AMERICA DO ABOUT GUN VIOLENCE?" JANUARY 30, 2013

It's an honor to be here today on behalf of more than 4.5 million moms and dads and sons and daughters, in every state across our nation, who make up the National Rifle Association of America. Those 4.5 million active members are joined by tens of millions of NRA supporters.

And it's on behalf of those millions of decent, hard-working, law-abiding citizens ... to give voice to their concerns ... that I'm here today.

The title of today's hearing is "What should America do about gun violence?" We believe the answer to that question is to be honest about what works—and what doesn't work.

Teaching safe and responsible gun ownership works—and the NRA has a long and proud history of teaching it.

Our "Eddie Eagle" children's safety program has taught over 25 million young children that if they see a gun, they should do four things: "Stop. Don't touch. Leave the area. Tell an adult." As a result of this and other private sector programs, fatal firearm accidents are at the lowest levels in more than 100 years.

The NRA has over 80,000 certified instructors who teach our military personnel, law enforcement officers and hundreds of thousands of other American men and women how to safely use firearms. We do more—and

spend more—than anyone else on teaching safe and responsible gun ownership.

We joined the nation in sorrow over the tragedy that occurred in Newtown, Connecticut. There is nothing more precious than our children. We have no more sacred duty than to protect our children and keep them safe. That's why we asked former Congressman and Undersecretary of Homeland Security, Asa Hutchinson, to bring in every expert available to develop a model School Shield Program—one that can be individually tailored to make our schools as safe as possible.

It's time to throw an immediate blanket of security around our children. About a third of our schools have armed security already—because it works. And that number is growing. Right now, state officials, local authorities and school districts in all 50 states are considering their own plans to protect children in their schools.

In addition, we need to enforce the thousands of gun laws that are currently on the books. Prosecuting criminals who misuse firearms works. Unfortunately, we've seen a dramatic collapse in federal gun prosecutions in recent years. Overall in 2011, federal weapons prosecutions per capita were down 35 percent from their peak in the previous administration. That means violent felons, gang members and the mentally ill who possess firearms are not being prosecuted. And that's unacceptable.

And out of more than 76,000 firearms purchases denied by the federal instant check system, only 62 were referred for prosecution and only 44 were actually prosecuted. Proposing more gun control laws—while failing to enforce the thousands we already have—is not a serious solution to reducing crime.

I think we can also agree that our mental health system is broken. We need to look at the full range of mental health issues, from early detection and treatment, to civil commitment laws, to privacy laws that needlessly prevent mental health records from being included in the National Instant Criminal Background Check System.

While we're ready to participate in a meaningful effort to solve these pressing problems, we must respectfully—but honestly and firmly—disagree with some members of this committee, many in the media, and all of the gun control groups on what will keep our kids and our streets safe.

Law-abiding gun owners will not accept blame for the acts of violent or deranged criminals. Nor do we believe the government should dictate what we can lawfully own and use to protect our families.

As I said earlier, we need to be honest about what works and what does not work. Proposals that would only serve to burden the law-abiding have failed in the past and will fail in the future.

Semi-automatic firearms have been around for over 100 years. They are among the most popular guns made for hunting, target shooting and self-defense. Despite this fact, Congress banned the manufacture and sale of hundreds of semi-automatic firearms and magazines from 1994 to 2004. Independent studies, including a study from the Clinton Justice Department, proved that ban had no impact on lowering crime.

And when it comes to the issue of background checks, let's be honest—background checks will never be "universal"—because criminals will never submit to them.

But there are things that can be done and we ask you to join with us. The NRA is made up of millions of Americans who support what works...the immediate protection for all—not just some—of our school children; swift, certain prosecution of criminals with guns; and fixing our broken mental health system.

We love our families and our country. We believe in our freedom. We're the millions of Americans from all walks of life who take responsibility for our own safety and protection as a God-given, fundamental right. (http://www.judiciary.senate.gov/imo/media/doc/1-30-13LaPierre Testimony.pdf)

EXCERPT OF WRITTEN TESTIMONY OF DAVID B. KOPEL, RESEARCH DIRECTOR OF THE INDEPENDENCE INSTITUTE AND ASSOCIATE POLICY ANALYST AT THE CATO INSTITUTE, BEFORE THE U.S. SENATE COMMITTEE ON THE JUDICIARY FOR THE HEARING "WHAT SHOULD AMERICA DO ABOUT GUN VIOLENCE?" JANUARY 30, 2013

Gun prohibition advocates have been pushing the "assault weapon" issue for a quarter century. Their political successes on the matter have always depended on public confusion. The guns are *not* machine guns. They do *not* fire automatically. They fire only one bullet each time the trigger is pressed, just like every other ordinary firearm. They are *not* more powerful than other firearms; to the contrary, their ammunition is typically intermediate in power, less powerful than guns and ammunition made for big game hunting.

The Difference between Automatic and Semi-automatic

For an automatic firearm (commonly called a "machine gun"), if the shooter presses the trigger and holds it, the gun will fire continuously, automatically, until the ammunition runs out. Ever since the National Firearms Act of 1934, automatics have been very strictly regulated by federal law: Every person who wishes to possess one must pay a $200 federal transfer tax, must be fingerprinted and photographed, and must complete a months-long registration process with the federal Bureau of Alcohol, Tobacco, Firearms, and Explosives (BATFE). In addition, the transferee must be granted written permission by local law enforcement, via ATF Form 4. Once registered, the gun may not be taken out of state without advance written permission from BATFE.

Since 1986, the manufacture of new automatics for sale to persons other than government agents has been forbidden by federal law. As a result, automatics in [the] U.S. are rare (there are about a hundred thousand legally registered ones), and expensive, with the least expensive ones costing nearly ten thousand dollars.

The automatic firearm was invented in 1883 by Hiram Maxim. The early Maxim Guns were heavy and bulky, and required a two-man crew to operate. In 1943, a new type of automatic was invented, the "assault rifle." The assault rifle is light enough for a soldier to carry for long periods of time. Soon, the assault rifle became the ubiquitous infantry weapon. Examples include the U.S. Army M-16, the Soviet AK-47, and the Swiss militia SIG SG 550. The AK-47 (and its various updates, such as the AK-74 and AKM) can be found all over the Third World, but there are only a few hundred in the United States, mostly belonging to firearms museums and wealthy collectors.

The precise definition of "assault rifle" is supplied by the Defense Intelligence Agency. If you use the term "assault rifle," persons who are knowledgeable about firearms will know precisely what kinds of guns you are talking about. The definition of "assault rifle" has never changed, because the definition describes a particular type of thing in the real world—just like the definitions of "apricot" or "Minnesota."

In contrast, the definition of "assault weapon" has never been stable. The phrase is merely an epithet. It has been applied to things which are not even firearms (namely, air guns). It has been applied to double-barreled shotguns, to single-shot guns (guns whose ammunition capacity is only a single round), and to many other sorts of ordinary handguns, shotguns, and rifles.

The first "assault weapon" ban in the United States, in California in 1989, was created by legislative staffers thumbing through a picture book of guns, and deciding which guns looked bad. The result was an incoherent law which, among other things, outlawed certain firearms that do not exist, since the staffers just copied the typographical errors from the book, or associated a model by one manufacturer with another manufacturer whose name appeared on the same page.

Over the last quarter century, the definition has always kept shifting. One recent version is Sen. Dianne Feinstein's new bill. Another is the pair of bills defeated in the January 2013 lame duck session of the Illinois legislature, which would have outlawed most handguns (and many long guns as well) by dubbing them "assault weapons."

While the definitions of what to ban keep changing, a few things remain consistent: The definitions do *not* cover automatic firearms, such as assault rifles. The definitions do *not* ban guns based on how fast they fire, or how powerful they are. Instead, the definitions are based on the name of a gun, or on whether a firearm has certain superficial accessories (such as a bayonet lug, or a grip in the "wrong" place).

Most, but not all, of the guns which have been labeled "assault weapons" are semi-automatics. Many people think that a gun which is "semi-automatic" must be essentially the same as an automatic. This is incorrect.

Semi-automatic firearms were invented in the 1890s, and have been common in the United States ever since. Today, about three-quarters of new handguns are semi-automatics. A large share of rifles and shotguns are also semi-automatics. Among the most popular semi-automatic firearms in the United States today are the Colt 1911 pistol (named for the year it was invented, and still considered one of the best self-defense handguns), the Ruger 10/22 rifle (which fires the low-powered .22 Long Rifle cartridge, popular for small game hunting or for target shooting at distances less than a hundred yards), the Remington 1100 shotgun (very popular for bird hunting and home defense), and the AR-15 rifle (popular for hunting game no larger than deer, for target shooting, and for defense). All of these guns were invented in the mid-1960s or earlier. All of them have, at various times, been characterized as "assault weapons."

Unlike an automatic firearm, a semi-automatic fires only one round of ammunition when the trigger is pressed. (A "round" is one unit of ammunition. For a rifle or handgun, a round has one bullet. For a shotgun, a single round contains several pellets).

In some other countries, a semi-automatic is usually called a "self-loading" gun. This accurately describes what makes the gun "semi"-automatic. When the gun is fired, the bullet (or shot pellets) travel from the firing chamber, down the barrel, and out the muzzle. Left behind in the firing chamber is the now empty case or shell that contained the bullets (or pellets) and the gunpowder.

In a semi-automatic, some of the energy from firing is used to eject the empty shell from the firing chamber, and then load a fresh round of ammunition into the firing chamber. Then, the gun is ready to shoot again, when the user is ready to press the trigger.

In some other types of firearms, the user must perform some action in order to eject the empty shell and load the next round. This could be moving a bolt back and forth (bolt action rifles), moving a lever down and then up (lever action rifles), or pulling and then pushing a pump or slide (pump action and slide action rifles and shotguns). A revolver (the second-most popular type of handgun) does not require the user to take any additional action in order to fire the next round.

The semi-automatic has two principle advantages over lever action, bolt action, slide action, and pump action guns. First, many hunters prefer it because the semi-automatic mechanism allows a faster second shot. The difference may be less than a second, but for a hunter, this can make all the difference.

Second, and more importantly, the semi-automatic's use of gunpowder energy to eject the empty case and then to load the next round substantially reduces how much recoil is felt by the shooter. This makes the gun much more comfortable to shoot, especially for beginners, or for persons without substantial upper body strength and bulk.

The reduced recoil also make[s] the gun easier to keep on target for the next shot, which is important for hunting and target shooting, and extremely important for self-defense.

Semi-automatics also have their disadvantages. They are much more prone to misfeeds and jams than are simpler, older types of firearms, such as revolvers or lever action.

Contrary to the hype of anti-gun advocates and less-responsible journalists, there is no rate of fire difference between a so-called "assault" semi-automatic gun and any other semi-automatic gun.

How Fast Does a Semi-automatic Fire?

Here is a report on the test-firing of a new rifle:

187 shots were fired in three minutes and thirty seconds and one full fifteen shot magazine was fired in only 10.8 seconds.

Does that sound like a machine gun? A "semi-automatic assault weapon"? Actually it is an 1862 test report of the then-new lever-action Henry rifle, manufactured by Winchester. If you have ever seen a Henry rifle, it was probably in the hands of someone at a cowboy re-enactment, using historic firearms from 150 years ago.

The Winchester Henry is a lever-action, meaning that after each shot, the user must pull out a lever, and then push it back in, in order to eject the empty shell casing, and then load a new round into the firing chamber.

The lever-action Winchester is not an automatic. It is not a semi-automatic. It was invented decades before either of those types of firearms. And yet that old-fashioned Henry lever action rifle can fire one bullet per second.

By comparison, the murderer at Sandy Hook fired 150 shots over a 20 minute period, before the police arrived. In other words, a rate of fewer than 8 shots per minute. This is a rate of fire far slower than the capabilities of a lever-action Henry Rifle from 1862, or a semi-automatic AR-15 rifle from 2010. Indeed, his rate of fire could have been far exceeded by a competent person using very old technology, such as a break-open double-barreled shotgun.

Are Semi-automatics More Powerful Than Other Guns?

The power of a firearm is measured by the kinetic energy it delivers. Kinetic energy is based on the mass (the weight) of the projectile, and its velocity. So a heavier bullet will deliver more kinetic energy than a lighter one. A faster bullet will deliver more kinetic energy than a slower bullet.

How much kinetic energy a gun will deliver has nothing to do with whether it is a semi-automatic, a lever action, a bolt action, a revolver, or whatever. What matter[s] is, first of all, the weight of the bullet, how much gunpowder is in the particular round of ammunition, and the length of the barrel.

None of this has anything to do with whether the gun is or is not a semi-automatic. Manufacturers typically produce the same gun in several different calibers, sometimes in more than a dozen calibers.

Regarding the rifles which some people call "assault weapons," they tend to be intermediate in power, as far as rifles go. Consider the AR-15 rifle in its most common caliber, the .223. The bullet is only a little bit wider than the puny .22 bullet, but it is longer, and thus heavier.

Using typical ammunition, an AR-15 in .223 would have 1,395 foot-pounds of kinetic energy. That's more than a tiny rifle cartridge like the .17 Remington, which might carry 801 foot-pounds of kinetic energy. In contrast,

a big-game cartridge, like the .444 Marlin, might have 3,040 [foot-pounds of kinetic energy]. This is why rifles like the AR-15 are suitable and often used for hunting small to medium animals (such as rabbits or deer), but are not suitable for the largest animals, such as elk or moose.

Many (but not all) of the ever-changing group of guns which are labeled "assault weapons" use detachable magazines (a box with an internal spring) to hold their ammunition. But this is a characteristic shared by many other firearms, including many non-semiautomatic rifles (particularly, bolt-actions), and by the large majority of handguns. Whatever the merits of restricting magazine size . . . , the size of the magazine depends on the size the magazine. If you want to control magazine size, there is no point in banning certain guns which can take detachable magazines, while not banning other guns which also take detachable magazines. (http://www.judiciary.senate.gov/imo/media/doc/1-30-13KopelTestimony.pdf)

EXCERPT OF TESTIMONY OF GAYLE S. TROTTER, SENIOR FELLOW AT THE INDEPENDENT WOMEN'S FORUM, BEFORE THE U.S. SENATE COMMITTEE ON THE JUDICIARY FOR THE HEARING "WHAT SHOULD AMERICA DO ABOUT GUN VIOLENCE?" JANUARY 30, 2013

I would like to begin with the compelling story of Sarah McKinley. Home alone with her baby, she called 911 when two violent intruders began to break down her front door. The men wanted to force their way into her home so they could steal the prescription medication of her deceased husband, who had recently died of cancer. Before the police could arrive, while Ms. McKinley was on the line with the 911 operator, these violent intruders broke down her door. One of the men brandished a foot-long hunting knife. As the intruders forced their way into her home, Ms. McKinley fired her weapon, fatally wounding one of the violent attackers and causing the other to flee the scene. Later, Ms. McKinley reflected on the incident: "It was either going to be him or my son," she said. "And it wasn't going to be my son."

Guns make women safer. Most violent offenders actually do not use firearms, which makes guns the great equalizer. In fact, over 90 percent of violent crimes occur without a firearm. Over the most recent decade, from 2001 to 2010, "about 6 percent to 9 percent of all violent victimizations were committed with firearms," according to a federal study. Violent criminals rarely use a gun to threaten or attack women. Attackers use their size and physical strength, preying on women who are at a severe disadvantage.

Guns reverse that balance of power in a violent confrontation. Armed with a gun, a woman can even have the advantage over a violent attacker. How do guns give women the advantage? An armed woman does not need superior strength or the proximity of a hand-to-hand struggle. She can protect her children, elderly relatives, herself or others who are vulnerable to an assailant. Using a firearm with a magazine holding more than 10 rounds of ammunition, a woman would have a fighting chance even against multiple attackers....

Concealed-carry laws reverse that balance of power even before a violent confrontation occurs. In this way, armed women indirectly benefit those who choose not to carry. For a would-be criminal, concealed-carry laws dramatically increase the cost of committing a crime, paying safety dividends to those who do not carry. All women in these jurisdictions reap the benefits of concealed-carry laws because potential assailants face a much higher risk when they attempt to threaten or harm a potential victim. As a result, in jurisdictions with concealed-carry laws, women are less likely to be raped, maimed or murdered than they are in states with stricter gun ownership laws.

Research has shown that states with nondiscretionary concealed handgun laws have 25 percent fewer rapes than states that restrict or forbid women from carrying concealed handguns. The most thorough analysis of concealed-carry laws and crime rates indicates that "there are large drops in overall violent crime, murder, rape, and aggravated assault that begin right after the right-to-carry laws have gone into effect" and that "in all those crime categories, the crime rates consistently stay much lower than they were before the law." Among the ten states that adopted concealed-carry laws over a fifteen-year span, there were 0.89 shooting deaths and injuries per 100,000 people, representing less than half the rate of 2.09 per 100,000 experienced in states without these laws.

Armed security works. Brave men and women stand guard over Capitol Hill, including the building where we are now. Snipers stand guard on the White House roof. Politicians and other high-profile individuals, including prominent gun-control advocates, have admitted to having gun permits either currently or in the past.

Armed guards often serve in the employ of those who themselves advocate for more restrictions on gun rights. Political figures seek to restrict gun rights, and Hollywood celebrities somberly urge Americans to "demand a plan" to reduce gun violence despite their own roles in graphically depicting lethal violence on the screen. In both cases, however, many of these political figures and celebrities already have their own plan: They rely on guns to safeguard their own personal safety. For example, armed guards protected a suburban newspaper in New York after the newspaper published the names and residential addresses of gun permit holders, and the newspaper's own reporter already used a gun for his protection. After publishing the story, the editors disclosed that their reporter owns a Smith & Wesson .357 Magnum and has "a residence permit in New York City."

While armed security works, gun bans do not. Anti-gun legislation keeps guns away from the sane and the law-abiding—but it does not keep guns out of the hands of criminals. Nearly all mass shootings have occurred in "gun-free" zones. Law-abiding citizens do not bring firearms to gun-free zones, so psychotic killers know they can inflict more harm in these unprotected environments. These laws make easy targets of the sane and the law-abiding. Gun-control advocates cheer the creation of legally mandated gun-free zones, touting increased safety while actually making citizens in those locations more vulnerable to the next horrible monster in search of soft targets. A moment's reflection confirms that statutory provisions and bold signs do not create a gun-free environment. No sober-minded person would advocate that approach when protecting banks, airports, rock concerts and government buildings. Instead, publicly designated gun-free zones have the effect of creating high-visibility soft targets—conspicuous environments where madmen can wreak havoc.

We need sensible enforcement of the gun laws that are already on the books. Currently, we have more than 20,000 under-enforced or selectively enforced gun laws. Gun regulation affects only the guns of the law-abiding. Criminals will not be bound by such gestures, especially as we continually fail to prosecute serious gun violations or provide meaningful and consistent penalties for violent felonies involving firearms.

Recently, a talk show host inadvertently exposed the absurdity of gun regulation in the District of Columbia when he displayed a 30-round magazine on national television, thereby embroiling himself in a police investigation. Ultimately, the Attorney General of the District of Columbia decided not to prosecute the matter. "Despite the clarity of the violation of this important law," he concluded, "a prosecution would not promote public safety." Why is it permissible to possess magazines to persuade people that guns are dangerous, but not for a woman to possess one to defend herself against gang rape? Overbroad anti-gun regulations unduly increase prosecutorial discretion and result in selective enforcement of the law. Equal justice under law should not depend on whether a prosecutor has a political or ideological motivation to seek enforcement. Nor should justice depend on whether a prosecutor has the good sense to decline enforcement of a knowing violation that does nothing more than unwittingly demonstrate the law's absurdity and overbreadth.

In lieu of empty, self-defeating gestures, we should address gun violence by doing what works. By safeguarding our Second Amendment rights, we preserve meaningful protection for women. Our nation made significant progress in that regard when, in recent memory, the United States Supreme Court held that the Second Amendment protects an individual's right to possess a firearm for traditionally lawful purposes, such as self-defense within the home.

For those who believe in safeguarding the civil liberties enshrined in our Bill of Rights, you might consider this an unremarkable conclusion. The constitutional text expressly guarantees the right "to keep and bear Arms," and that right is specifically enumerated—not implied—and guaranteed to "the people." In other words, unlike many of the individual rights that the Supreme Court has recognized—some would say invented—you can actually find the right to bear arms in the literal text of the Second Amendment. Moreover, the Constitution guarantees a "right of the people" only two other times, both of which clearly describe individual rights: The First Amendment protects the "right of the people" to assemble and to petition the government, and the Fourth Amendment protects the "right of the people" against "unreasonable searches and seizures."

Even so, dissenting liberals decried "the Court's announcement of a new constitutional right to own and use firearms for private purposes." Ironically, this claim originated from those who agree with the judicial philosophy that has discovered new fundamental individual rights hiding within "penumbras" that are "formed by emanations" from "specific guarantees in the Bill of Rights." Adherents to this view maintain that the Bill of Rights generates "penumbral emanations" that create assorted individual rights. However, they simultaneously claim that enforcing an individual right expressly written in the black letter of the constitutional text is the "announcement of a new constitutional right." On the one hand, shadowy secretions reveal the hidden meaning of rights secretly embedded in the Constitution and awaiting judicial divination. On the other hand, they view a specifically enumerated guarantee in the Bill of Rights as "a new constitutional right."

Moreover, the dissenting justices claimed that a local law could ban private possession of any form of [an] operable firearm because "the adjacent states do permit the use of handguns for target practice, and those states are only a brief subway ride away." They called this a "minimal burden" on the Second Amendment right to bear arms, as if a law-abiding citizen who is facing down an attacker might somehow have the ability to coax him onto the subway and take a brief ride to the adjoining jurisdiction's nearest target range. Adherents of this judicial philosophy—which purports to allow the restriction of individual liberties as long as "a brief subway ride" would transport an aggrieved citizen to another jurisdiction where the penumbral emanations flow freely—would assuredly provide more robust protection for rights of their own judicial invention.

These are two dramatically different views of our Bill of Rights. One approach has repeatedly created new rights found nowhere in the Constitution while unflinchingly limiting the Second Amendment's "right of the people to keep and bear Arms" to protect only the right to have a gun in the army, as peculiar as that would be. The other approach, which has twice prevailed in the Supreme Court, takes seriously the people's enumerated rights—the ones actually written in the Constitution—and respects the Second Amendment.

In lieu of empty gestures, we should address gun violence based on what works. Guns make women safer. The Supreme Court has recognized that lawful self-defense is a central component of the Second Amendment's guarantee of the right to keep and bear arms. For women, the ability to arm ourselves for our protection is even more consequential than for men because guns are the great equalizer in a violent confrontation. As a result, we preserve meaningful protection for women by safeguarding our Second Amendment rights. Every woman deserves a fighting chance. (http://www.judiciary.senate .gov/imo/media/doc/1-30-13TrotterTestimony.pdf)

APPENDIX: STATE CONSTITUTION ARTICLES CONCERNING WEAPONS

The Second Amendment of the Bill of Rights of the U.S. Constitution reads: "A well regulated militia, being necessary to the security of a free state, the right of the people to keep and bear arms, shall not be infringed."

Alabama: That every citizen has a right to bear arms in defense of himself and the state. Art. I, § 26 (enacted 1819).

Alaska: A well-regulated militia being necessary to the security of a free state, the right of the people to keep and bear arms shall not be infringed. The individual right to keep and bear arms shall not be denied or infringed by the State or a political subdivision of the State. Art. I, § 19 (first sentence was enacted in 1959; second sentence was added in 1994).

Arizona: The right of the individual citizen to bear arms in defense of himself or the State shall not be impaired, but nothing in this section shall be construed as authorizing individuals or corporations to organize, maintain, or employ an armed body of men. Art. II, § 26 (enacted in 1912).

Arkansas: The citizens of this State shall have the right to keep and bear arms, for their common defense. Art. II, § 5 (enacted in 1874).

California: No constitutional provision.

Colorado: The right of no person to keep and bear arms in defense of his home, person and property, or in aid of the civil power when thereto legally summoned, shall be called in question; but nothing herein contained shall be construed to justify the practice of carrying concealed weapons. Art. II, § 13 (enacted in 1876).

Connecticut: Every citizen has a right to bear arms in defense of himself and the state. Art. I, § 15 (enacted in 1818).

Delaware: A person has the right to keep and bear arms for the defense of self, family, home and State, and for hunting and recreational use. Art. I, § 20 (enacted in 1987).

Florida: (a) The right of the people to keep and bear arms in defense of themselves and of the lawful authority of the state shall not be infringed, except that the manner of bearing arms may be regulated by law. (b) There shall be a mandatory period of three days, excluding weekends and legal holidays, between the purchase and delivery at retail of any handgun. For the purposes of this section, "purchase" means the transfer of money or other valuable consideration to the retailer, and "handgun" means a firearm capable of being carried and used by one hand, such as a pistol or revolver. Holders of a concealed weapon permit as prescribed in Florida law shall not be subject to the provisions of this paragraph. (c) The legislature shall enact legislation implementing subsection (b) of this section, effective no later than December 31, 1991, which shall provide that anyone violating the provisions of subsection (b) shall be guilty of a felony. (d) This restriction shall not apply to a trade in of another handgun. Art. I, § 8 (sections (b) to (d) adopted in 1990).

Georgia: The right of the people to keep and bear arms shall not be infringed, but the General Assembly shall have power to prescribe the manner in which arms may be borne. Art. I, § 1, Paragraph VIII (enacted in 1877).

Hawaii: A well regulated militia being necessary to the security of a free state, the right of the people to keep and bear arms shall not be infringed. Art. I, § 17 (enacted in 1959).

Idaho: The people have the right to keep and bear arms, which right shall not be abridged; but this provision shall not prevent the passage of laws to govern the carrying of weapons concealed on the person nor prevent passage of legislation providing minimum sentences for crimes committed while in possession of a firearm, nor prevent the passage of legislation providing penalties for the possession of firearms by a convicted felon, nor prevent the passage of any legislation punishing the use of a firearm. No law shall impose licensure, registration or special taxation on the ownership or possession of firearms or ammunition. Nor shall any law permit the confiscation of firearms, except those actually used in the commission of a felony. Art. I, § 11 (enacted in 1890).

Illinois: Subject only to the police power, the right of the individual citizen to keep and bear arms shall not be infringed. Art. I, § 22 (enacted in 1970).

Indiana: The people shall have a right to bear arms, for the defense of themselves and the State. Art. I, § 32 (enacted in 1851).

Iowa: No constitutional provision.

Kansas: The people have the right to bear arms for their defense and security; but standing armies, in time of peace, are dangerous to liberty, and shall not be tolerated, and the military shall be in strict subordination to the civil power. Bill of Rights, § 4 (enacted in 1859).

Kentucky: All men are, by nature, free and equal, and have certain inherent and inalienable rights, among which may be reckoned: ... The right to bear arms in defense of themselves and of the State, subject to the power of the General Assembly to enact laws to prevent persons from carrying concealed weapons. Bill of Rights, § 1 (enacted in 1891).

Louisiana: The right of each citizen to keep and bear arms shall not be abridged, but this provision shall not prevent the passage of laws to prohibit the carrying of weapons concealed on the person. Art. I, § 11 (enacted in 1974).

Maine: Every citizen has a right to keep and bear arms and this right shall never be questioned. Art. I, § 16 (enacted in 1820).

Maryland: No constitutional provision.

Massachusetts: The people have a right to keep and to bear arms for the common defence [*sic*]. And as, in time of peace, armies are dangerous to liberty, they ought not to be maintained without the consent of the legislature; and the military power shall always be held in an exact subordination to the civil authority, and be governed by it. Pt. I, Art. XVII (enacted in 1780).

Michigan: Every person has a right to keep and bear arms for the defense of himself and the state. Art. I, § 6 (enacted in 1963).

Minnesota: No constitutional provision.

Mississippi: The right of every citizen to keep and bear arms in defense of his home, person, or property, or in aid of the civil power when thereto legally summoned, shall not be called in question, but the legislature may

regulate or forbid carrying concealed weapons. Art. III, § 12 (enacted in 1890).

Missouri: That the right of every citizen to keep and bear arms in defense of his home, person and property, or when lawfully summoned in aid of the civil power, shall not be questioned; but this shall not justify the wearing of concealed weapons. Art. I, § 23 (enacted in 1875).

Montana: The right of any person to keep or bear arms in defense of his own home, person, and property, or in aid of the civil power when thereto legally summoned, shall not be called in question, but nothing herein contained shall be held to permit the carrying of concealed weapons. Art. II, § 12 (enacted in 1889).

Nebraska: All persons are by nature free and independent, and have certain inherent and inalienable rights; among these are life, liberty, the pursuit of happiness, and the right to keep and bear arms for security or defense of self, family, home, and others, and for lawful common defense, hunting, recreational use, and all other lawful purposes, and such rights shall not be denied or infringed by the state or any subdivision thereof. To secure these rights, and the protection of property, governments are instituted among people, deriving their just powers from the consent of the governed. Bill of Rights, Art. I, § 1 (right to keep and bear arms enacted in 1988).

Nevada: Every citizen has the right to keep and bear arms for security and defense, for lawful hunting and recreational use and for other lawful purposes. Art. I, § 11(1) (enacted in 1982).

New Hampshire: All persons have the right to keep and bear arms in defense of themselves, their families, their property and the state. Pt. I, Art. IIa (enacted in 1982).

New Jersey: No constitutional provision.

New Mexico: No law shall abridge the right of the citizen to keep and bear arms for security and defense, for lawful hunting and recreational use and for other lawful purposes, but nothing herein shall be held to permit the carrying of concealed weapons. No municipality or county shall regulate, in any way, an incident of the right to keep and bear arms. Art. II, § 6 (first sentence enacted in 1971; second sentence added in 1986).

New York: No constitutional provision.

North Carolina: A well regulated militia being necessary to the security of a free State, the right of the people to keep and bear arms shall not be infringed; and, as standing armies in time of peace are dangerous to liberty, they shall not be maintained, and the military shall be kept under strict subordination to, and governed by, the civil power. Nothing herein shall justify the practice of carrying concealed weapons, or prevent the General Assembly from enacting penal statutes against that practice. Art. I, § 30 (enacted in 1971).

North Dakota: All individuals are by nature equally free and independent and have certain inalienable rights, among which are those of enjoying and defending life and liberty; acquiring, possessing and protecting property and reputation; pursuing and obtaining safety and happiness; and to keep and bear arms for the defense of their person, family, property, and the state, and for lawful hunting, recreational, and other lawful purposes, which shall not be infringed. Art. I, § 1 (right to keep and bear arms enacted in 1984).

Ohio: The people have the right to bear arms for their defense and security; but standing armies, in time of peace, are dangerous to liberty, and shall not be kept up; and the military shall be in strict subordination to the civil power. Art. I, § 4 (enacted in 1851).

Oklahoma: The right of a citizen to keep and bear arms in defense of his home, person, or property, or in aid of the civil power, when thereunto legally summoned, shall never be prohibited; but nothing herein contained shall prevent the Legislature from regulating the carrying of weapons. Art. II, § 26 (enacted in 1907).

Oregon: The people shall have the right to bear arms for the defence [*sic*] of themselves, and the State, but the Military shall be kept in strict subordination to the civil power. Art. I, § 27 (enacted in 1857).

Pennsylvania: The right of the citizens to bear arms in defense of themselves and the State shall not be questioned. Art. I, § 21 (enacted in 1790).

Rhode Island: The right of the people to keep and bear arms shall not be infringed. Art. I, § 22 (enacted in 1843).

South Carolina: A well regulated militia being necessary to the security of a free State, the right of the people to keep and bear arms shall not be infringed. As, in times of peace, armies are dangerous to liberty, they shall not be maintained without the consent of the General Assembly. The military power of the State shall always be held in subordination to the civil authority and be governed by it. Art. I, § 20 (enacted in 1895).

South Dakota: The right of the citizens to bear arms in defense of themselves and the state shall not be denied. Art. VI, § 24 (enacted in 1889).

Tennessee: That the citizens of this State have a right to keep and to bear arms for their common defense; but

the Legislature shall have power, by law, to regulate the wearing of arms with a view to prevent crime. Art. I, § 26 (enacted in 1870).

Texas: Every citizen shall have the right to keep and bear arms in the lawful defense of himself or the State; but the Legislature shall have power, by law, to regulate the wearing of arms, with a view to prevent crime. Art. I, § 23 (enacted in 1876).

Utah: The individual right of the people to keep and bear arms for security and defense of self, family, others, property, or the state, as well as for other lawful purposes shall not be infringed; but nothing herein shall prevent the Legislature from defining the lawful use of arms. Art. I, § 6 (enacted in 1984).

Vermont: That the people have a right to bear arms for the defence [*sic*] of themselves and the State—and as standing armies in time of peace are dangerous to liberty, they ought not to be kept up; and that the military should be kept under strict subordination to and governed by the civil power. Ch. I, Art. XVI (enacted in 1777).

Virginia: That a well regulated militia, composed of the body of the people, trained to arms, is the proper, natural, and safe defense of a free state, therefore, the right of the people to keep and bear arms shall not be infringed; that standing armies, in time of peace, should be avoided as dangerous to liberty; and that in all cases the military should be under strict subordination to, and governed by, the civil power. Art. I, § 13 (enacted in 1971).

Washington: The right of the individual citizen to bear arms in defense of himself, or the state, shall not be impaired, but nothing in this section shall be construed as authorizing individuals or corporations to organize, maintain or employ an armed body of men. Art. I, § 24 (enacted in 1889).

West Virginia: A person has the right to keep and bear arms for the defense of self, family, home and state, and for lawful hunting and recreational use. Art. III, § 22 (enacted in 1986).

Wisconsin: The people have the right to keep and bear arms for security, defense, hunting, recreation or any other lawful purpose. Art. I, § 25 (enacted in 1998).

Wyoming: The right of citizens to bear arms in defense of themselves and of the state shall not be denied. Art. I, § 24 (enacted in 1889).

IMPORTANT NAMES
AND ADDRESSES

Americans for Responsible Solutions
PO Box 15642
Washington, DC 20003
E-mail:
info@americansforresponsiblesolutions.org
URL: http://www.americansforresponsible
solutions.org/

**Brady Campaign to Prevent Gun
Violence**
840 First St. NE, Ste. 400
Washington, DC 20002
(202) 370-8100
URL: http://www.bradycampaign.org/

**Bureau of Alcohol, Tobacco,
Firearms, and Explosives**
**Office of Public and Governmental
Affairs**
99 New York Ave. NE, Rm. 5S 144
Washington, DC 20226
(202) 648-8500
URL: http://www.atf.gov/

**Centers for Disease Control
and Prevention**
1600 Clifton Rd.
Atlanta, GA 30329-4027
1-800-232-4636
URL: http://www.cdc.gov/

Coalition to Stop Gun Violence
805 15th St. NW, Ste. 700
Washington, DC 20005
(202) 408-0061
E-mail: csgv@csgv.org
URL: http://www.csgv.org/

Federal Bureau of Investigation
935 Pennsylvania Ave. NW
Washington, DC 20535-0001
(202) 324-3000
URL: http://www.fbi.gov/

Gun Owners of America
8001 Forbes Place, Ste. 102
Springfield, VA 22151
(703) 321-8585
FAX: (703) 321-8408
URL: http://www.gunowners.org/

Law Center to Prevent Gun Violence
268 Bush Street, Ste. 555
San Francisco, CA 94104
(415) 433-2062
FAX: (415) 433-3357
URL: http://smartgunlaws.org/

National Institute of Justice
810 Seventh St. NW
Washington, DC 20531

(202) 307-2942
URL: http://www.ojp.usdoj.gov/nij

**National Rifle Association
of America**
11250 Waples Mill Rd.
Fairfax, VA 22030
1-800-672-3888
URL: http://www.nra.org/

**Office of Juvenile Justice and
Delinquency Prevention**
810 Seventh St. NW
Washington, DC 20531
(202) 307-5911
URL: http://www.ojjdp.gov/

**Second Amendment
Foundation**
12500 NE 10th Place
Bellevue, WA 98005
(425) 454-7012
E-mail: info@saf.org
URL: http://www.saf.org/

Violence Policy Center
1730 Rhode Island Ave. NW, Ste. 1014
Washington, DC 20036
(202) 822-8200
URL: http://www.vpc.org/

RESOURCES

The Bureau of Alcohol, Tobacco, Firearms, and Explosives (ATF) monitors the production and regulation of alcohol, tobacco, firearms, and explosives and is the major source for statistical and technical information on these categories. In the annual *State Laws and Published Ordinances—Firearms*, the ATF provides a complete overview of firearms regulations of towns, cities, states, and the federal government. The ATF also publishes the annual report *Firearms Commerce in the United States*, which supplies authoritative data on the manufacturing, import/export, and sales of firearms.

The Bureau of Justice Statistics (BJS) and the Federal Bureau of Investigation (FBI) maintain statistics on crime in the United States. The FBI's annual *Crime in the United States* is based on crime statistics reported by law enforcement agencies through its Uniform Crime Reports program. The FBI also publishes the annual *Law Enforcement Officers Killed and Assaulted*. The BJS supplements this information with data derived from surveys of victims and other supplementary sources. Among the most useful BJS publications in the preparation of this book were *Homicide in the U.S. Known to Law Enforcement, 2011* (Erica L. Smith and Alexia Cooper, December 2013), *Firearm Violence, 1993–2011* (Michael Planty and Jennifer L. Truman, May 2013), and *Criminal Victimization, 2012* (Jennifer Truman, Lynn Langton, and Michael Planty, October 2013). Information on the National Instant Criminal Background Check System is available online at http://www.fbi.gov/hq/cjisd/nics.htm.

Other important government data came from the Centers for Disease Control and Prevention, which operates the Web-based Injury Statistics Query and Reporting System (http://www.cdc.gov/injury/wisqars/). It also publishes the annual *Health, United States*, which provides data on mortality and injuries in addition to many other health-related topics, and periodic reports relevant to the topic of firearm violence as part of its National Violent Death Reporting System in the *Morbidity and Mortality Weekly Report*.

Gale, Cengage Learning would like to express its continuing appreciation to the Gallup Organization for its kind permission to publish its surveys and to the National Opinion Research Center of the University of Chicago for use of its material. Thanks also goes to the Pew Research Center and Harris Interactive for permission to publish material from their surveys. Gale, Cengage Learning would also like to thank the following organizations for permission to use information and tables from their web postings, reports, and journals: the National Rifle Association of America and the Brady Campaign to Prevent Gun Violence.

INDEX

Bureau of Justice Statistics (BJS)
  on homicide rate, 58
  Homicide Trends in the United States, 1
  NCVS of, 57
Bush, George W.
  NICS Improvement Amendments Act, 31
  Protection of Lawful Commerce in Arms Act, 56
Butler, Robert, Jr., 114
Butterfield, Fox, 56
Byrne, Sklar v., 52

C

Caldwell, Cornell, 56
California
  crime guns recovered in, 106
  crime guns, tracing, 79
  FFLs in, 31
  firearm homicides in, 62
  gun manufacturer immunity law, 56
  Hoosier v. Randa, 53
  NFA weapons registered in, 11
  open carry prohibited in, 36
  Paula Fiscal et al. v. City and County of San Francisco et al., 50
  Saturday Night Specials ban, 53
  schoolyard shooting in Stockton, 12, 29
  shooting at Oikos University, Oakland, 115
  Silveira v. Lockyer, 42
  waiting period requirement in, 32
California Court of Appeals, 53
California Supreme Court, 50
CAP. *See* Child access prevention (CAP) laws
Case, Curtis, 114
Cases v. United States, 41
Cases Velazquez, Jose, 41
"Causes of Law Enforcement Deaths" (National Law Enforcement Officers Memorial Fund), 68
CDC. *See* Centers for Disease Control and Prevention
CDF (Children's Defense Fund), 116–117
Centers for Disease Control and Prevention (CDC)
  data on gun-related injuries/fatalities, 83
  funding for prevention of gun violence, 133
  on nonfatal gunshot injuries, 83–84
  on student acquisition of firearms, 106
"Changes in Suicide Rates by Hanging and/or Suffocation and Firearms among Young Persons Aged 10–24 Years in the United States: 1992–2006" (Bridge et al.), 95
Chardon High School, Chardon, OH, 114
Charles II, King of England, 2
Chicago City Council, 52

Chicago, Illinois
  gun traces in, 106
  limits on handgun possession, 52–53
  McDonald v. Chicago, 5, 47
Chicago, McDonald v., 5, 47
Chicago Office of the Mayor, 106
Child access prevention (CAP) laws
  AAP on responsible gun ownership, 134
  in Connecticut, 48–49
  effects of, 94–95
"Child Access Prevention Policy Summary" (Law Center to Prevent Gun Violence), 49, 94
Children
  AAP on gun control laws and, 133–134
  BB/pellet gun injuries among children aged 0 to 19 years, rates per 100,000, 86(t6.4)
  death for young people aged 1 to 24 years, leading causes of, 95f
  deaths from gun violence, 131
  firearm deaths of children aged 0 to 19 years, rates per 100,000, 98t
  firearm fatalities in rural/urban areas, 90
  gun ownership in households with, 123
  gun safety programs to protect, 95
  guns in the home, risks of, 15
  homicide by firearm, likelihood of, 98
  homicide firearm deaths of children aged 12/under, rate per 100,000, 102t
  homicide victims, 100
  injured/killed by gunfire, 116–117
  nonfatal firearm injuries among, 84, 86(t6.3)
  nonfatal firearm violence rate for, 72
  safe storage of guns in home, 94–95
  Sandy Hook Elementary School shooting, 125
  See also School shootings; Youth
Children's Defense Fund (CDF), 116–117
Chili's (restaurant chain), 36
Chipotle (restaurant chain), 36
Cho, Seung-Hui
  NICS Improvement Amendments Act, 32
  shooting at Virginia Polytechnic Institute and State University, 66, 114–115
Cicero, Marcus Tullius, 1
Circumstances
  law enforcement officers accidentally killed, by circumstance at scene of incident, 70t
  law enforcement officers assaulted, by circumstance at scene of incident/type of weapon, 71(t5.12)
  law enforcement officers killed, by circumstance at scene of incident, 69(t5.9)
  of police deaths/injuries, 68
Cities
  gun violence in, 131–132
  nonfatal firearm violence rates for, 72
  youth gangs in, 105

City and County of San Francisco et al., Paula Fiscal et al. v., 50
City of Baltimore, Barron v., 39
City of Renton, Second Amendment Foundation v., 50–51
Coalition of New Jersey Sportsmen v. Whitman, 49
Coker, Billy Wayne, 54
Coker, Wal-Mart Stores, Inc. v., 54
Collective rights
  individual rights *vs.*, 5, 39–40
  United States v. Emerson and, 42
Colonies, American, 3
Colorado
  Aurora movie theater shooting, 76
  Denver's right to ban certain guns, 53
  shooting at Columbine High School, 66, 113–114, 121
  shooting at Deer Creek Middle School, 116
  shooting at Platte Canyon High School, 115
  state constitution article, 143
Colt 1911 pistol, 139
Columbine High School, Littleton, CO
  description of shooting at, 113–114
  mass shooting at, 66
  public debate over gun control, 121
The Columbine High School Shootings: Jefferson County Sheriff Department's Investigation Report (Jefferson County Sheriff Department), 113
Commentaries (Blackstone), 2
Commerce clause
  Gun-Free School Zones Act and, 44, 45
  United States v. Stewart, 46
Concealed weapons
  concealed-carry laws, 141
  permits, 47
  state laws on, 32–34
"Concord Hymn" (Emerson), 2
Connecticut
  Benjamin v. Bailey, 48
  concealed weapons laws, 32–34
  open carry, permit for, 36
  shooting at Sandy Hook Elementary School, 97, 116, 125, 127
  state constitution article, 143
  State v. Wilchinski, 48–49
Connecticut Supreme Court
  Benjamin v. Bailey, 48
  State v. Wilchinski, 49
ConocoPhillips, 49
Constitution Society, 3
Cook, Philip J., 25
Cooper, Alexia
  on firearms used in homicides, 61–62
  on homicide demographics, 58, 61
  on homicide rate, 58
  on youth homicides, 97, 101–103

CPSIA information can be obtained
at www.ICGtesting.com
Printed in the USA
FFOW05n1404030615

9 781573 026420